AF651876

BENGAL DISTRICT GAZETTEERS.

DARJEELING.

BENGAL DISTRICT GAZETTEERS

DARJEELING

L S S O'MALLEY

LOGOS PRESS
New Delhi-110002

BENGAL DISTRICT GAZETTEERS

Burdwan/**J.C.K. Paterson**
Darjeeling/**L.S.S. O' Malley**
Howrah/**L.S.S O' Malley**
Hooghly/**L.S.S. O' Malley**
Santal Parganas/L.S.S **O' Malley**

ISBN 81-7268-018-X

First Published 1907

Reprinted 1985
Second Reprint 1999

Published by
LOGOS PRESS
4788-90/23, Ansari Road,
Darya Ganj,
New Delhi
Phone : 3272187

Printed at
Tarun Offset Printers
Delhi

BENGAL DISTRICT GAZETTEERS.

DARJEELING.

BY

L. S. S. O'MALLEY,

INDIAN CIVIL SERVICE.

THE BENGAL SECRETARIAT BOOK DEPÔT.

1907

PREFACE.

I DESIRE to acknowledge my indebtedness to Mr. C. A. Bell's Report on the Survey and Settlement of the Kalimpong Government Estate (1905), which I have found of great assistance in compiling this volume. I beg also to express my obligations to the Rev. R. Kilgour, B. D., Mr. F. A. Möller and Mr. Claud Bald for their contributions.

L. S. S. O'MALLEY.

PLAN OF CONTENTS.

TABLE OF CONTENTS.

CHAPTER I.

PHYSICAL ASPECTS.

CHAPTER II.

HISTORY.

CHAPTER III.

THE PEOPLE.

CHAPTER IV.

PUBLIC HEALTH.

CHAPTER V.

AGRICULTURE.

CHAPTER VI.

THE TEA INDUSTRY.

CHAPTER VII.

FORESTS.

CHAPTER VIII.

NATURAL CALAMITIES.

CHAPTER IX.

RENTS, WAGES AND PRICES.

CHAPTER X.

OCCUPATIONS, MANUFACTURES AND TRADE.

CHAPTER XI.

MEANS OF COMMUNICATION.

CHAPTER XII.

LAND REVENUE ADMINISTRATION.

CHAPTER XIII.

GENERAL ADMINISTRATION.

CHAPTER XIV.

LOCAL SELF-GOVERNMENT.

CHAPTER XV.

EDUCATION.

CHAPTER XVI.

GAZETTEER.

GAZETTEER

OF THE

DARJEELING DISTRICT.

CHAPTER I.

PHYSICAL ASPECTS.

GENERAL DESCRIPTION. THE district of Darjeeling lies between 26° 31′ and 27° 13′ north latitude, and between 87° 59′ and 88° 53′ east longitude. It contains a total area of 1,164 square miles, and a population of 249,117 souls. The principal town, which is also the administrative head-quarters of the district, is Darjeeling, situated in the lower Himālayas in 27° 3′ north latitude and 88° 16′ east longitude. The name Darjeeling is a corruption of *dorje*, the precious stone or ecclesiastical sceptre, which is emblematic of the thunderbolt of Sakhra (Indra) and of *ling*, a place. It means therefore the place of the *dorje*, the mystic thunderbolt of the Lāmaist religion, this being the name by which the Buddhist monastery which once stood on Observatory Hill was formerly known.

Boundaries. In shape the district resembles an irregular triangle, of which the apex projects into British territory, while the base rests on Sikkim. It is a frontier district, running up between Nepāl and Bhutān and stretching from the plains of Bengal on the south to the State of Sikkim on the north. The British frontier is demarcated from the latter State by a series of rivers and mountain torrents, from Nepāl on the west by the lofty Singalīlā chain of mountains, and from Bhutān on the north-east by the Jaldhākā river, while its south-eastern and southern boundary marches with the British districts of Jalpaiguri and Purnea.

Natural configuration. The territory comprised within the district is not marked out by any natural features as a region complete in itself. It comprises an area which, as a portion of the great Himalayan range, is quite insignificant, and the limits of which have been determined by political considerations. It is no larger than a small English county, but it ranges in altitude from 300 to 12,000 feet above sea-level. The Tarai stretches along the base of the hills, a low-lying strip shut in on the north by the giant outliers of the Himālayas. Geographically, this tract belongs to the plains of India, but geologically it is a sort of neutral country, being composed in places neither of the alluvium of the plains nor of the rocks of the hills, but, for a great part, of alternating beds of sand, gravel and boulders brought down from the mountains. Botanically, it is readily defined as the region of forest trees, among which the *sāl* (*Shorea robusta*) is conspicuous. The remainder of the district consists of a mass of mountainous spurs and ranges rising to the height of 12,000 feet. There are no flat valleys or plains in the whole country, no lakes or precipices of any consequence below that elevation, and few or no bare slopes, except where the virgin forest has yielded place to the tea-garden or the field of the cultivator. The main ranges wind and zig-zag in all directions, giving off a number of long spurs on either flank; the valleys thus formed present a great variety of climate and elevation. For the most part, they stretch from north to south, while the courses of the principal rivers are in the same direction; but many of the spurs and of the torrents flowing between them run east and west, and even in some cases from south to north. Consequently, the interior of the district has more than once been described as a confused labyrinth of ridges and valleys.

Natural Divisions. From the preceding account it will be seen that Darjeeling falls naturally into two distinct tracts, the Tarai immediately beneath the hills, and the ridges and deep valleys of the lower Himālayas. The Tarai portion of the district is a low-lying belt of country, traversed by numerous rivers and streams rushing down from the hills and by the upland ridges which mark their courses. It is an unhealthy marshy tract, formerly covered by dense malarious jungle, in which aboriginal tribes of Meches, Dhimāls and Koches burnt clearings and raised their scanty crops of rice and cotton on a system, if system it can be called, of nomadic husbandry. It has now been extensively cleared for tea-gardens and settled tillage, but still contains large blocks of *sāl* forest interspersed with cultivated land and village sites. Behind the Tarai, the mountains tower abruptly from the plains, which are here only about 300 feet above sea-level, in lofty spurs reaching to 6,000 and 10,000 feet, and culminate in a series of long ridges and valleys. According to one of the earliest travellers in the district, the mountains in 1830 were completely clothed with forest from the top to the very bottom, and formed rather a sombre feature in the landscape, owing to the sameness of tint and want of break or variety on the surface. The slopes, from about 6,000 feet downwards, are now dotted with trim tea-gardens, interspersed with small tracts of land reserved for native cultivators. Above that level they are clothed in dense forest, through which torrents rush down, their position often only indicated by the dipping of the forest into their beds; but at the higher levels on the Singalīlā range there are wide grassy slopes broken here and there with pine forests and masses of rhododendrons.

There is a great difference between the scenery presented by this part of the district and that met with in the belt of country skirting the foot of the hills. The more elevated portion of the Tarai contains numerous tea-gardens, the monotonous fertility of which affords a striking contrast to the extensive forests elsewhere. Lower down, the country is more open and is dotted with cultivators' homesteads, enclosed in shady groves of bamboo and plantain. The general nature of the scenery in this part of Darjeeling has been well described in Sir Joseph Hooker's* account of the view from Pankhābāri. "Behind, the Himālayas rise in steep confused masses. Below, the ranges, as far as the eye can reach east and west, throw spurs on to the plains of India. These are very thickly wooded, and enclose broad, dead-flat, hot and damp valleys. The Tarai district forms a very irregular belt, scantily clothed, and intersected by innumerable rivulets from the hills, which unite and divide again on the flat, till, emerging from the region of many trees, they enter the plains, following devious courses, which glisten like silver threads. The whole horizon is bounded by the sea-like expanse of the plains, which stretch away into the region of sunshine and fine weather, in one boundless flat." Scenery.

To the north of the Tarai, the Himālayas stand out in a succession of bold spurs, the appearance of which has been compared with that of the weather-beaten front of a mountainous coast. In the interior the scenery is of a wilder and more magnificent description. Here the hills form a series of lofty ridges containing deep valleys, many thousand feet deep, and still covered in many places with a dense mass of forest trees festooned with moss and dripping with moisture. These shady forests give the scenery of Darjeeling a distinctive character, as they contain a very luxuriant vegetation, ranging from the matted cane-brakes of the Tarai and the tropical forms of the lower valleys to the oak

* Himalayan Journals.

and pine forests which clothe the highest ridges. Along the banks of the larger streams, and from 1,500 to 2,000 feet up the mountain ridges, is a tropical forest which also extends across the level flats in the Tarai at the base of the outer spurs. On dry slopes in the hills and along old river beds in the Tarai there is little undergrowth, but elsewhere it is dense and luxuriant. In the forests which clothe the hills in the sub-temperate zone from 2,000 to 5,000 feet, bamboos, plantains and palms are numerous, the screw pine is commonly seen, and the graceful tree-fern with its feathery crown is abundant. Above 5,000 feet is the temperate zone, in which maples, oaks, chestnuts and magnolias are met with in great abundance; the trees are mantled with ferns, mosses and epiphytic orchids; and the white-flowered magnolia blossoms so profusely in places that the forests on the flanks of the mountains appear as if sprinkled with snow. Orchids cease to be abundant at 8,000 feet, from which level rhododendrons are plentiful, some growing into great trees and others clothing the hill-sides with a thick growth of shrubs. At a still higher level the ridges are crowned by clumps of pine trees, often blasted by lightning or twisted and gnarled by the violent storms to which they are exposed. The effect in either case is equally picturesque, as they stand out against the line of snowy peaks which bounds the horizon.

MOUNTAINS. The Himalayan range has been commonly divided into three zones:—the great range of snowy peaks, which, roughly speaking, form the axis of the chain; the Lower or Outer Himālayas, forming a broad belt of mountains of inferior, though still considerable, altitude south of the snows; and, thirdly, the comparatively low hills forming the Sub-Himalayan zone, either as ridges or spurs contiguous with the outer hills or separated from them by the flat-bottomed valleys known as *Duns*. In Darjeeling the mountains belong to the Lower Himālaya zone and consist of long tortuous ranges, running generally from north to south throughout its length. The Sub-Himalayan zone is altogether wanting, and the detached ridges met elsewhere in the Himālayas are unrepresented; while the snowy range lies far beyond the limits of the district to the north, where it gives the appearance of a long range of mountains stretching east and west at an average distance of about 50 miles. This range forms the great backbone of the Darjeeling Himālayas. To the north-west tower the giant peaks of Kinchinjunga (28,146 feet), and to the north east is Dongkya (23,184 feet) at a distance of rather less than 50 miles. From Kinchinjunga the Singalīlā range, an immense ridge 60 miles long, stretches south to the plains, forming the boundary between Nepāl and Darjeeling. It is the continuation of this ridge in a south and then south-easterly direction by Tanglu and Senchal, with its various lateral spurs, which constitutes the Darjeeling hill territory west of the Tīsta. To the east of that river a lofty ridge runs southwards from Dongkya, dividing at Gipmochi (11,518 feet) into two great spurs, one of which runs to the south-east and the other to the south-west, including between them the valley of the Jaldhākā. It is the lower half of the south-western spur, with its numerous ramifications, that constitutes the hills of Kālimpong east of the Tīsta. The highest point of these hills is where the main ridge first enters British territory, where it has an altitude of over 10,000 feet; the other eminences do not exceed 7,000 or 8,000 feet.

The Singalilā ridge is the most interesting one in this part of the Himālayas. It commands Nepāl on one side, and Sikkim and Darjeeling on the other; and it is the water-shed of two great river systems, for the streams on its western flank run into the Kosi, an affluent of the Ganges, while those on its eastern flank run into the Tīsta, an affluent of the Brahmaputra. It is a ridge ranging from 10,000 to 12,000 feet high, the highest peaks of which are Sandakphu (11,929 feet), and Phalūt (11,811 feet), where the boundaries of Nepāl, Darjeeling and Sikkim meet. A number of spurs descend eastward from the main ridge into the Darjeeling territory. These spurs are of great extent, one being 6 miles long, two 12 miles, and many 2 or 4 miles in length.

South-east of the Singalīlā range four great hill ranges radiate from a central point at Ghum, a saddle 7,372 feet in elevation situated to the north-west of Senchal. One, the Ghum range, climbs due west to Simāna-bastī, where it meets the Singalīlā range, the elevation of the crest varying from 7,000 to 7,900 feet. To the south, the Senchal-Mahaldiram range trends generally southwards towards Kurseong, reaching an elevation of 8,000 to 8,600 feet in its northern half and about 7,000 feet towards the south, the highest points being East Senchal (8,600 feet), Tiger Hill (8,515 feet) and West Senchal (8,163 feet). The third range, the Takdāh or Tangbu range, branches off from the main Senchal ridge below Tiger Hill, at a saddle 3 miles east of Ghum, and then takes a north-easterly direction, sinking gradually from the height of 7,300 feet to 680 feet at the junction of the Great Rangīt and Tīsta rivers. The fourth range, the Darjeeling-Jalāpahār range, on which are situated the cantonments of Katāpahār and Jalāpahār and the civil station of Darjeeling, extends northward from Ghum, at first rising rather abruptly to the height of 7,886 feet, then gradually descending to 7,002 feet at the Chaurāsta in Darjeeling, and again rising to 7,163 feet at Observatory Hill. At this point the range divides into two, the Lebong spur and the Takvār spur, enclosing between them the narrow valley of the Rangīt, a tributary of the Great Rangīt.

To the east of the Tīsta, the country is cut up by ridges of varying height and steepness, separated by narrow valleys, the largest of which run far back into the mountains. These ridges debouch into the plains at 300 to 1,000 feet above sea-level, rising in the west to 10,500 feet at Rishi-lā, a mountain not for from the trijunction point of Darjeeling with Sikkim and Bhutān. Other prominent peaks in this portion of the district are Khempung (7,800 feet) and Songchonglu (6,300 feet), both on the outer range overlooking the Duārs, and Padamlu (6,900 feet) and Damsong (6,300 feet) on the inner range rising above the Tīsta.

RIVER SYSTEM. The valleys on the south-eastern side of the Singalīlā ridge are drained by the Mechī, Bālasan and Mahānadī rivers; and all the remainder of the Darjeeling hills is drained by the Tīsta and its tributaries, except the extreme eastern end, where the chief effluent is the Jaldhākā. A short account of the principal rivers is given below.

The Tīsta. The Tīsta, like many of the other great rivers of Northern India, rises on the further side of the Himālayas and bursts through the mountain barrier before it reaches British territory. After draining Sikkim, the Tīsta forms the boundary between that State and the Darjeeling district for some distance, till it receives the waters of the Great Rangīt. It then runs to the south, and, threading its way through the mountains, debouches on the plains through a gorge known as the Sivok Golā Pass. After a short course through the Darjeeling Tarai, it passes into the Jalpaigurī district at its north-western corner; and eventually falls into the Brahmaputra in the Rañgpur district, 168 miles distant from the point at which it enters this district. In Darjeeling its principal tributaries are the Rangpo and the Rillī on its left bank, and the Great Rangīt, the Rangpo, the Rayeng, and the Sivok on its right. A little below its junction with the Great Rangīt, where one of the most picturesque views in the district can be obtained, a fine suspension bridge has been thrown across the stream. This bridge plays an important part in the communications of the district, as it connects Darjeeling with the great trade route across the Jelep pass into Tibet.

The Tīsta is not navigable in its course through the hills, where it is a broad mountain torrent, with numerous shallows and rapids. Its current is very swift, running in places at the rate of 14 miles an hour, and it is liable to sudden rises. When it enters the plains, it has a width of 700 or 800 yards, and becomes navigable by small boats, but for some distance navigation is very difficult and precarious, owing to the rapids and the numerous rocks and boulders in the bed of the river. In the dry season its waters are sea-green; but after rain the admixture of calcareous detritus gives them a milky hue. The scenery along the river banks is extremely beautiful. The lower slopes of the mountains are clothed in dense forest overhanging its waters, which rush down over a rocky bed between high banks, now shelving and now abrupt, while in the background rise, tier above tier, the great snowy masses of the Himālayas.

The Great Rangīt, the chief affluent of the Tīsta, enters Darjeeling district from the west and forms part of the northern boundary, flowing from west to east till it joins the Tīsta. Its affluents, above the point of junction with the Tīsta, are the Rangnu, the Little Rangīt, and the Rammān, which successively fall into it upon its southern bank. Above the point where it receives the waters of the Rammān, the course of the Great Rangīt lies entirely in Sikkim. The river, being purely a mountain stream, is not navigable within Darjeeling at any time of the year. It has shelving banks, generally clothed with forest, but with patches of cultivation here and there, and a stony and sandy bed. One of the most picturesque places along its course is where it meets the waters of the Tīsta. Here there is a great difference in the colour of the two rivers, the Tīsta being turgid and milky white, the Rangīt dark green and very clear. The waters preserve their colour for some hundred yards, and the line separating the two is distinctly marked. There is no less marked a difference in the temperature of the two rivers, as the water of the Rangīt is appreciably warmer than that of the Tīsta. The colour and coldness of the latter are no doubt due to the number of glaciers which it drains; while the Rangīt is chiefly supplied by the rainfall of the outer ranges of the Senchal and Singalīlā hills, and hence its water is warmer and clearer, except during the height of the rains. The Great Rangīt.

The Rammān, one of the tributaries of the Great Rangīt, takes its rise under the Phalūt peak in the Singalīlā range, which forms the western boundary of the district. It first touches on Darjeeling in the extreme north-west of the district, and then flowing from west to east forms the boundary between Darjeeling and Sikkim until it falls into the Great Rangīt. The principal tributaries of the Rammān within this district are the Rathu The Rammān.

and Siri rivers, which take their rise within the district, and, flowing northwards, empty themselves into the Rammān on its right bank. The latter is crossed by a curious natural bridge of stone between the points where the Rathu and Siri rivers join it.

The Little Rangīt. The Little Rangīt takes its rise under the Tanglu mountain in the Singalilā range on the borders of Nepāl, and flows generally in a north-easterly direction till it falls into the Great Rangīt on its southern bank.

The Rangnu. The Rangnu takes its rise under Senchal, flows northwards past the station of Darjeeling, and empties itself into the right bank of the Great Rangīt. It is practically a mountain torrent which comes tearing down from Senchal, several thousand feet above its junction with the Rangīt; and though its roar is heard, and its course is visible throughout its length, the stream itself cannot be seen clearly from above, so deep does it cut into its channel.

The Mahānadī. The Mahānadī has its source near Mahaldiram to the east of Kurseong. After leaving the hills it flows in a southerly direction as far as Silīgurī, where it changes its course a little to the west and forms the boundary-line between the Tarai and Jalpaigurī as far as Phānsidewā in the extreme south-east of the district. After leaving Darjeeling, the Mahānadī passes through Purnea and Mālda, and finally falls into the Ganges within the Rājshāhi district. The river receives no tributaries of any importance within the limits of Darjeeling, and only attains its full volume after leaving the district. The name Mahānadī, or as it is also called the Mahānanda, is a Bengali corruption of Mahaldi, the Lepchā name for this river.

The Bālasan. The Bālasan river takes its rise at Lepchā Jagat a few miles to the south-west of the station of Darjeeling. It flows a southerly course till soon after it enters the Tarai, when it divides into two streams. One, called the New Bālasan, branches off and joins the Mahānadī on its right bank just below Silīgurī; the other, the Old Bālasan, continues its southward course till passing out of the Tarai, it also joins the Mahānadī in Purnea. The new channel is said to have been formed about 60 years ago by some Meches damming up the old stream for the purpose of fishing. The principal tributaries of the Bālasan in the hills are the Rinchingtong on the left, and the Rangbang on the right bank; and in the plains the Rakti and Rohini, both of which join it on its left bank.

The Mechī. The Mechī takes its rise under the Rangbang spur in the Singalīlā range on the Nepāl frontier, and flowing from north to south marks the western boundary of the district from its source. After it enters the Tarai, it divides into two branches near the lower Mechī forest, and eventually joins the Mahānadī in the Purnea district.

The Rillī, a tributary of the Tīsta, is the most important stream in the tract situated to the east of the Tīsta. It takes its rise under the Khampang mountain in the north-east of the district, and flows in a winding south-westerly course till it falls into the Tīsta. The Rilli

The Jaldhākā, called in the upper part of its course the Di-chhu, marks the eastern boundary of the district, which it separates from Bhutān and from the Western Duārs in the Jalpaigurī district. It runs a straight course from north to south, its principal tributaries being the Paralangchhu, Rangchhu and Machhu, which flow into it on its right bank. The Jaldhākā.

The rocks of the Darjeeling district were subdivided by Mr. Mallet into five groups, viz., gneiss, the Dāling series, the Buxa series, Gondwānas, and the Tertiary system. The outcrops of these form a series of bands, running more or less parallel to the general trend of the Himālayas and dipping one beneath the other into the hills. The most curious feature of these subdivisions is that the younger formations always appear to underlie the older:—thus the Tertiary beds disappear under the Gondwānas, the Gondwānas under the Buxa and Dāling series, and the latter under the gneiss, the original order of superposition having been completely reversed by folding and faulting. GEOLOGY.*

The gneiss varies from a foliated granitoid rock composed of quartz, felspar and biotite to a more or less pure mica-schist, and include partly intrusive granite and partly metamorphosed beds of sedimentary origin. The Dāling series covers a large area in the northern and eastern parts of the district. It consists of phyllite, slate and quartzite with some hornblende-schist and very subordinate bands of dolomite and crystalline limestone. Copper ore is frequently found disseminated through the slates and schists. The Buxa series, which is largely developed in the Western Duārs, occurs only at the extreme eastern end of Darjeeling district. It consists of slates, quartzites and dolomite, the predominance of the latter rock serving as a means of distinction from the Dāling series. The Gondwāna beds crop out near the base of the hills and constitute a narrow band between the Dālings and the Tertiaries, running from Pankhābāri to Dālingkot. They consist chiefly of sandstone, shale and coal, all of which have been intensely

* The account of the Geology of the Darjeeling district has been contributed by Mr. H. H. Hayden, Superintendent, Geological Survey of India.

crushed and faulted, and dip at high angles to the north-north-west; they are frequently metamorphosed by pressure, and their component rocks converted into quartzites, slates and graphitic schists, in which case it is difficult to distinguish them from the Dāling series. Owing to this crushing, the coal-seams vary much in thickness and are occasionally cut out altogether by faults. The Tertiary beds fringe the older rocks continuously from close to the Mechī eastward nearly as far as Dālingkot. They are chiefly composed of soft, massive, "pepper and salt" sandstones, containing mica and felspar, with clunchy grey micaceous and calcareous beds containing a few subordinate layers of limestone. The sandstone frequently contains lignite, which, however, has not been found in sufficient quantity to be of economic value.

Coal occurs in the band of Gondwāna rocks which runs from near Pankhābāri to Dālingkot. The beds usually dip at high angles to the north-north-west, but are much contorted and faulted, and the coal is frequently badly crushed. The seams were examined first by Mr. Mallet, and subsequently by Mr. Bose, who found them to contain a considerable amount of coking coal of good quality, having an ash content sometimes only a little over 12 per cent. From 1896 to 1900 attempts were made to work the seams, but the result was not, apparently, satisfactory, the difficulty of getting the coal and the crushed nature of most of the seams probably acting as a bar to profitable exploitation. Very impure graphite occurs in the semi-graphitic Gondwāna schists of the Rakti river, but is of no economic value. Iron, in the form of hematite and ferruginous clay, is found at Lohārgarh, and high grade magnetite and micaceous hematite, free from sulphur and phosphorus, forms a band about 20 feet thick at Sikbhar. Copper ore occurs disseminated through the rocks of the Dāling series at very large number of places in the district. The ore, which is usually chalcopyrite, has long been known to the natives, who mine it in their usual primitive fashion and smelt it with charcoal. No attempt has yet been made to work on a large scale or by modern methods, the few localities that have from time to time been prospected by Europeans, having been always abandoned eventually. There are three possible sources of lime in the district, viz., the dolomite of the Buxa series, the limestone bands in the Tertiary rocks, and calcareous tufa. Neither the dolomite nor the limestone are used, however, probably owing to the great purity of the tufa, which contains over 98 per cent. of carbonate of lime, and is obtainable at numerous localities, being deposited by springs which issue chiefly along the line of junction of the Gondwānas with the Tertiary rocks.*

The vegetation of the Darjeeling district is peculiarly rich in the number of species and peculiarly varied in its character, as might be expected from a consideration of its climate and physiography. Broadly speaking, the vegetation of the district shows two well-marked zones—a tropical zone from the plains up to about 6,000 feet, and a temperate zone from about 6,000 to 12,000 feet. BOTANY.†

The lowest part of the tropical zone comprising the slopes leading up to the base of the outer hills is characterized, especially in its western half, by forests of *Shorea robusta* (*sāl*), with a mixture of other trees such as *Dillenia pentagyna*, *Butea frondosa*, and species of *Terminalia* and *Eugenia*. Here and there also occur large stretches of Savannah forest of such species as *Dillenia pentagyna*, *Butea frondosa*, *Eugenia obovata*, etc. Along the river banks *Dalbergia Sissoo* is fairly common, accompanied by *Acacia Catechu*, *Bombax malabaricum*, *Nauclea cordifolia*, *Garuga pinnata*, etc. Patches of mixed forest also occur formed by a great number of species, of which the more common are *Schima Wallichii*, *Terminalia tomentosa*, *Terminalia myriopteron*, *Artocarpus Chaplasha*, *Bombax malabaricum*, *Dillenia indica*, *Eugenia formosa* and many species of figs. This mixed forest also extends up the valleys and lower spurs to about 3,000 feet, sometimes with *sāl* predominating, although other very common trees are *Cedrela*, *Duabanga*, *Careya*, *Lagerstrœmia*, *Magnolia*, *Michelia*, and many *Leguminosæ*. At higher elevations still, up to about the limit of the tropical zone, this forest still continues, but its general character is altered by the more frequent appearance of such species as *Engelhardtia spicata*, *Castanopsis indica*, *Cerasus Puddum*, *Alnus nepalensis*, *Bucklandia populnea*, *Juglans regia* (the walnut), oaks and maples. Several species of palms occur in the tropical zone, such as *Areca*, *Wallichia*, *Phœnix rupicola*, *Caryota urens*, *Plectocomia*, and climbing *Calami*. Screw pines, one true pine (*Pinus longifolia*), various species of bamboo of the genera *Dendrocalamus*

---

* For further details *see* The Geology of the Darjeeling district, etc., Mem., Geol. Surv. India, XI (1875), by F. R. Mallet; The Darjeeling coal between the Lisu and Ramthi rivers, Records, Geol. Surv. India, XXIII, p. 237 (1890), by P. N. Bose; Further note on the Darjeeling Coal Exploration, Records, Geol. Surv. India, XXIV, p. 212 (1891), by P. N. Bose; The Geology and Mineral Resources of Sikkim, Records, Geol. Surv. India, XXIV, p. 217 (1891), by P. N. Bose, and, on a deposit of copper ore near Komai, Darjeeling district. Records, Geol. Surv. India, XXXI, cap. (1904), by H. H. Hayden.

† I am indebted to Captain A. T. Gage, I.M.S., Superintendent of the Royal Botanical Garden, Calcutta, for this account of the Botany of Darjeeling.

Arundinaria, etc., are quite common. The shrubby element in the vegetation is made up by *Acanthaceæ*, *Melastomaceæ*, *Rubiaceæ*, *Urticaceæ*, etc. Of climbers there are many species of *Ampelideæ*, *Cucurbitaceæ*, *Convolvulaceæ*, *Apocynaceæ*, *Asclepiadaceæ*, *Smilax*, *Dioscorea*, *Rhaphidophora*, the latter being especially conspicuous. The herbaceous vegetation is well represented by *Malvaceæ*, balsams, *Polygonaceæ*, orchids, *Scitamineæ*, and *Aroideæ* with many other widely distributed species. Tree ferns are fairly common, and other ferns of such genera as *Davallia*, *Pteris*, *Asplenium*, *Nephrodium*, *Polypodium*, *Angiopteris*, *Gleichenia*, etc., abound. In swampy places tall grasses of the genera *Saccharum*, *Thysanolæna*, *Erianthus*, *Anthistiria* are very common. In cleared spots, a scrub of *Artemisia* and *Leucosceptrum canum* is very common.

The temperate zone is characterized by forests of gigantic trees, consisting chiefly of oaks, chestnuts, magnolias, Michelias, laurels, maples, birches, and on the Singalīlā ridge conifers. Of the shrubby vegetation, the chief species belong to *Clematis*, *Berberis*, *Ilex*, *Rosa*, *Rubus*, *Contoneaster*, *Spiræa*, *Aucuba*, *Lonicera*, *Osbeckia*, *Vaccinium*, *Pentapterygium*, *Ericaceæ*, *Polygonum*, etc. A few rhododendrons grow within the altitude of the greater part of the district, but the rhododendron forests are found only on the high points above 8,000 feet along the Singalīlā ridge. Dwarf bamboos abound. The herbaceous vegetation is composed of such plants as aconites, violets, many balsams, *Potentillas*, *Fragarias*, *Chrysosplenium*, gentians, *Campanulas*, lilies, etc.

FAUNA.* Although the Darjeeling district is small in size, it contains a rich variety of fauna, owing to the great difference between the climatic conditions of its northern and southern parts, the former extending to a height of 12,000 feet in the Himālayas, and the latter stretching down to the plains. Among the larger Carnivora, the leopard (*Felis pardus*) is found in this northern tract, and occasionally the ounce (*Felis uncia*), the clouded leopard (*Felis Diardi*) and lynx (*Felis Isabilina*), but these are only stray visitors from higher altitudes. In the Tarai the tiger (*Felis tigris*) and leopard (*Felis pardus*) are common; while the large tiger-cat (*Felis viverrina*), the marbled tiger-cat (*Felis marmorata*), the leopard-cat (*Felis bengalensis*) and the common jungle-cat (*Felis chaus*) have their habitat in both tracts. The tiger is met with in fairly large numbers in the plains portion of the district, as well as in the lower hills up to the height of 2,000 feet; and it is said to have been found, in a few instances, as high as 7,000 feet. Man-eaters are scarce, and it is probable

* This account of the Fauna of the district has been prepared from an article kindly supplied by Mr. F. A. Möller of Darjeeling.

that in the few cases in which human beings are killed by tigers, they are not killed for the sake of food, as the place is so well stocked with game and cattle that tigers have no need to take to human flesh. Leopards are numerous in the valleys throughout the district, where they do much damage by carrying off sheep, goats, pigs and dogs from the villages. They are easily caught in traps, with a goat or a dog as bait; and during the last few years some 40 leopards have been trapped in this way at Tindhāria on the Darjeeling-Himalayan Railway by one of the railway employés. The principal species of Viverra are the large civet-cat (*Viverra zibetha*), the lesser civet-cat (*Viverra malaccensis*), the tiger civet (*Prinodon pardicolor*), and several species of the Paradoxura and mongoose family. The jackal (*Canis aureus*) is the only representative of the genus Canis; the wild dog (*Cyon rutilans*) of the genus Cuon; and the Indian fox (*Vulpis bengalensis*) and hill fox (*Vulpis montanus*) of the genus Vulpis. Wild dogs are rare, and are seldom met with, though a pack of them caused a good deal of damage in 1905 to cattle in the direction of Laba.

The order Ungulata comprise the elephant, the wild pig and various Ruminantia. The elephant (*Elephas indicus*) is found in the large forest tracts at the foot of the hills up to 3,000 feet, but is known to travel up to the height of 10,000 feet. There are at present three main herds of wild elephants in the district. One comes up from the Tendu forest in Jalpaigurī, and usually goes as far as the Naksal Khar, a large salt-lick on a tributary of the Jaldhākā river to the east of the Kālimpong hills; but they also go further into the hills reaching an altitude of 10,000 feet on the Rishi-lā and even roam about in the snow at that height. Another herd, about 20 in number, works fairly constantly between the Tīsta and Jaldhākā rivers, their head-quarters being the Māl and East Nar forest blocks, though they also ascend the hills and graze in the higher tracts adjoining Bhutān; while a third herd, containing at least 30 animals, ranges from the Tīsta westwards across the Mechī river into Nepāl. The result is that the whole belt of country to the south of the district is infested by wild elephants, sweeping through it in large herds or roaming singly. To the east of the Tīsta they confine their ravages for the most part, to the produce of the forests, but to the west of that river they are a veritable scourge to the people; and for the last few years the Tarai has had a melancholy record of persons killed, crops destroyed, and villages ruined by them. Indeed, the depredations of these animals in this part of the district have become so serious a menace to life and property that there is a danger of much of the land being thrown out of cultivation and relapsing into jungle. The wild pig (*Sus indicus*) is found throughout the district up to 8,000 feet.

The Ruminantia include the *gaur*, the deer tribe and goats. The *gaur* (*Bos gaurus*) is found in the same tract of country as the elephant, where they are fairly numerous. They have been preserved for the last few years, but this measure is perhaps unnecessary, for it would probably be sufficient to lay down that only bulls may be killed. These brutes are dangerous, especially solitary ones, which are very vicious, as a rule, and occasionally attack travellers. Of the deer tribe the *sāmbar* (*Rusa aristotelis*) is found in the forest tracts at the foot of the hills up to 3,000 feet, and before the district was opened out for tea and for native cultivation, it was found in all the large *sāl* forests among the larger rivers, as far up as the Great Rangīt river. The spotted deer (*Axis maculatus*) is only found west of the Bālasan river, but a stray one has been known to cross to the east. The hog-deer (*Axis porcinus*) was formerly plentiful in the Tarai, but it is now scarce having been almost exterminated, partly owing to the extension of cultivation in the Tarai, and partly in consequence of the prevalence of netting, in which old and young, buck and doe, are killed indiscriminately. It is partial to grass lands, is hardly ever found in the forests, and never goes up into the hills. The barking-deer (*Cervulus aureus*) is found over the whole of the district. The goat tribe comprise the *serow* (*Nemorhædus bubalina*) and the *gorāl* (*Nemorhædus goral*).

Among the Ursidæ the Himalayan black bear (*Ursus torquatus vel Tibetanus*), and the common Indian sloth bear (*Ursus labiatus*) are common. The Malayan sun-bear (*Ursus Malayanus*) has also once been met with. The bear first named is plentiful in the hills up to 10,000 feet, and is also found at the foot of the hills. It is often described as being very fierce, vicious and liable to attack any one it comes across without provocation. Mr. Möller states that this certainly applies to those found in the plains; indeed, one could hardly pass a Mechī village 20 years ago without finding one or more of the villagers horribly mutilated, invariably in the face, and in every instance, they had been attacked by bears without any provocation. This description does not, however, apply to the hill species, as it is very seldom that any one is attacked by them. On the other hand, they do considerable damage to the maizē crops in the hills; and in 1904 a bear established itself near Sonāda, and killed and partially ate several cows. The genus Lepus include the common Indian red-tailed hare (*Lepus ruficaudatus*) and the hispid hare (*Lepus hispidus*); the latter however is very rare.

Other mammalia are monkeys, squirrels (of which the two Himalayan flying squirrels are worthy of note) and several species of porcupines, martens, moles, civets, rats and mice.

In the northern portion of the district the moonal pheasant (*Lophophorus impeyanus*), tragopan (*Ceriornis satyra*), blood pheasant (*Ithagenes cruentus*), *kalij* pheasant (*Euplocamus leucomelanus*), the common wood partridge (*Arboricola torqueola*), the rufous-throated wood-partridge (*Arboricola rufogularis*), wood-cock (*Scolopax rusticola*) and red jungle-fowl (*Gallus ferrugineus*) are met with; and in the southern part, the lesser florican (*Sypheotides bengalensis*), red jungle-fowl, red spur-fowl (*Galloperdix spadiceus*), pea-fowl (*Pavo cristatus*), *kalij* and black partridge (*Francolinus vulgaris*), besides several species of quail, snipe, duck and waders. Game birds.

There are about 47 species of snakes found in the district, of which 17 are more or less poisonous. The following species are most frequently found:—the hamadryad or king cobra (*Naia bungarus*), the common cobra (*Naia tripudians*), the *karait* (*Bungarus cæruleus*), the banded *karait* (*Bungarus fasciatus*), one viper (*Vipera Russellii*), and several of the genus Trimeresurus. The python is found in the Tarai and the lower valleys. The long-nosed fish-eating crocodile has been seen in the Mahānadī river, but it is rare. Snakes.

The fish found in the district include the mahseer, the *katli*, the fresh-water shark and many others. The mahseer and the *katli* give good sport, the former running up to 50 pounds. Before the cyclone of September 1899, the Tīsta, the Great and Little Rangīt, and the Rammān rivers were well stocked with fish, and local sportsmen had a good time with the rod. But it appears that nearly all the big fish were either killed or carried off to sea, for very few have been taken since that time; and it is only now, after the lapse of 7 years, that the rivers seem to have recovered. Fish.

Owing to the tropical situation of the town of Darjeeling, the seasons largely follow the course of those in the plains, the cold weather, hot weather and rains, but there are two short periods—and those the most delightful in the year—which correspond in some ways to an English autumn and spring. The cold weather is divided into two portions. The first at the end of rains is mild and pleasant, the atmosphere being tolerably clear, and generally free from mist and cloud. This is the autumn, if autumn there be, at Darjeeling. Towards the beginning of December, the first touch of winter comes with hoar-frost; and at the end of that month and in January, the ground is sometimes frozen almost the whole day. The air is cloudless, dry and bracing; in the early morning CLIMATE.

it is very cold, but later in the day there is bright sunshine, and it becomes pleasantly warm, though it remains bitterly cold in the shade. As the evening comes on, it is cold and chilly, and the nights are clear and starry. Occasionally snow falls in January and February, but such an occurrence is comparatively rare, and heavy falls of snow are very uncommon. It seldom lies on the ground for more than a few hours, except in places sheltered from the sun; but in January 1883 it lay on the ground for 10 days, and in February 1887 for three weeks and more on the higher ranges near Darjeeling. From the accounts of early travellers it would seem that in Darjeeling itself snow was more frequent formerly. In February 1829, when General Lloyd first visited the place, snow fell for three successive days in quantities sufficient to cover Darjeeling and the whole of the surrounding heights; and when he revisited it again in January 1837, snow covered the ground to the depth of a foot or more, some of it remaining unthawed for over a week.

In March a brief Himalayan spring is ushered in with high blustering winds; it is very short, lasting only till the end of the month, but during this time rhododendrons, magnolias, primroses and other garden flowers come out in bloom, and the slopes lose the brown dried-up appearance they get in winter. During April and May there is a short-lived summer accompanied by showers of rain, which become heavier and more frequent till the setting in of the rains in the beginning of June. For three months after this, Darjeeling is exposed to the full force of the monsoon, drenched with rain, and shrouded in mist. The alluvial plain between it and the mouths of the Ganges is almost a dead level, the foot of the hills being only 300 feet above the sea; and consequently the vapour-laden southerly winds from the Bay of Bengal reach the outer range of hills without impediment. The humidity is very great, and Darjeeling is at this period of the year one of the dampest stations in India, while the neighbouring ranges which catch the full force of the monsoon are enveloped in mist and clouds, which rarely lift. The rains continue unabated during the months of June, July and August, 24 inches on the average falling in the first, 32 inches in the second and 26 inches in the last of those three months. The rainfall varies moreover very greatly in different parts of the station; in the present year (1906) it amounted by the end of August to 86 inches at North Point (6,513 feet), while it was 103 inches at the Government Observatory near St. Paul's School (7,376 feet) and was as much as 117 inches at the Municipal Office in the bazar. With September a decided change takes place, the continued rainfall giving way to showers, which become less and less frequent, while the sun shows itself oftener and its gleams are of longer duration, till towards the latter end of the month or the beginning of October, when the rains cease altogether.

During this period of the year the weather, even when apparently most settled, cannot be depended upon for any time, and Darjeeling is subject to almost constant cloud and fog, which rise from the deep humid valleys and hang for days together over the station. The station is shut in by a screen of mountains which, while sheltering it from wind, prevent the dissipation of the rolling masses of mists, which rise from the ravines like steam from a cauldron. Owing to this natural barrier, the winds are mostly light and variable; and for a considerable portion of the year they have not sufficient strength to dispel the canopy of clouds lying on the hills which surround the town. On the average, there are two days of clouded sky to one day of clear sky.

As the Darjeeling district consists of valleys, in which the elevation above the sea may be not more than 1,000 feet, and hill ranges, which rise to a height of 10,000 to 12,000 feet, climatic conditions are necessarily varied, more especially as regards temperature. In the Tarai and the lower valleys the heat is tropical; but in the town of Darjeeling the temperature is that of a temperate zone and only about two degrees above that of London.* At the station itself, where the observatory stands at a height of 7,376 feet, the mean temperature is about 42° in the cold-weather months from December to February. A rapid increase of temperature takes place during March and April owing to the warmer air which penetrates through the valleys from the plains; and thereafter, from May to December, when more or less heav cloudy prevails, mean temperature is steady at or slightly above 60°. In October northerly winds begin, cloud is much less than in previous months, and rainfall occurs mainly owing to cyclonic storms which generally recurve towards North Bengal at the end of the season. The lowest average minimum temperature is 35° in January and the highest 58° in July. Temperature.

During the cold-weather months very little rain falls in the Darjeeling hills. November and December are almost rainless, and the light showers which fall in January and February occur Rainfall.

* The following are the average annual temperatures at London and Darjeeling:—

	London.	Darjeeling.
Average maximum temperature ...	57·6	58·5
„ minimum „ ...	42·7	47·5
„ air „ ...	50·2	52·7

when shallow depressions are passing eastward over the plains. The heaviest rainfall recorded in January is 3 inches in 1899, and in Febuary 2·65 inches in 1889. Rainfall is somewhat more heavy in March, and there is a considerable increase in April, owing to thunderstorms. In May the southerly winds from the Bay reach the hills and cause increased precipitation, which is at times very heavy, more especially in the latter half of that month, when cyclonic storms pass from the Bay into Bengal. Very heavy rainfall is general from June to September, especially on the lower slopes. The monsoon current flows northwards over East Bengal; and in North Bengal receives its westward deflection towards North-Western India. At the same time, there is great ascensional motion, maintained as far as the spurs increase in height, and to that ascensional motion the heavy rainfall is due. In the wettest month (July) Mangpu at 4,700 feet receives 32 inches, Kurseong at about the same level 45 inches, and Darjeeling 32 inches. The total fall at these stations during the monsoon season, *i.e.*, from May to November, is 116, 153, and 114 inches respectively. The monsoon rainfall at Silīgurī in the Tarai is 116 inches; at Pedong, at a height of over 6,000 feet, it is 90 inches; while on the lower hills to the extreme east it is about 200 inches. Statistics of the rainfall at the various recording stations are given below for the cold weather (November to February), the hot weather (March to May), and the rainy season (June to October). The figures shewn are the averages recorded from the earliest year in which rainfall was systematically registered up to the end of 1905: –

Station.	Years recorded.	November to February.	March to May.	June to October.	Annual average.
Darjeeling	40-43	2·28	13·92	105·60	121·80
Kālimpong ...	17-19	2·14	10·77	76·61	89·52
Kurseong	17-18	2·39	18·23	140·10	160·72
Mangpu	8	2·09	16·30	105·66	124·05
Pedong	7-8	2·99	17·30	83·14	103·43
Siliguri	17-18	1·57	15·35	105·55	122·47

CHAPTER II.

HISTORY.

THE history of Darjeeling presents a late chapter in the extension of British rule, for it was not until the beginning of the 19th century that the East India Company was brought into direct relations with the tract of country which now bears the name. It then formed part of the dominions of the Rājā of Sikkim, a petty ruler who had long been engaged in an unsuccessful struggle against the growing power of the warlike Gurkhas. After overrunning the hills and valleys of Nepāl, they marched east into Sikkim in 1780; and during the next 30 years the country suffered repeatedly from their inroads. At the end of this period, they had overrun Sikkim as far eastward as the Tīsta river, and had conquered and annexed the Tarai, *i.e.*, the belt of country lying along the lower hills between that river and the Mechī, which is now covered by the valuable tea-gardens of the Darjeeling planters. In the meantime, the East India Company was engaged in unavailing remonstrances against the Nepalese aggressions throughout the whole length of their northern frontier, and war finally broke out in 1814. At its close, the tract which the Nepalese had wrested from the Rājā of Sikkim was ceded to the East India Company; the Rājā, who had been driven out of his dominions, was reinstated; and in 1817 a treaty was concluded at Titālya, under which the whole of the country between the Mechī and the Tīsta, a tract extending over 4,000 square miles, was restored to him, his sovereignty being guaranteed by the Company. The intervention of the British was thus successful in preventing the Gurkhas from turning the whole of Sikkim and the hills west and south of the Tīsta into an outlying province of Nepāl; and Sikkim, including the present district of Darjeeling, was retained as a buffer state between Nepāl and Bhutān.

THE NEPALESE WAR.

Under this treaty, the Company assumed the position of the paramount power in Sikkim, the Rājā being bound to refer to the arbitration of the British Government any disputes between his subjects and those of Nepāl or any other neighbouring State. Ten years after it was signed, disputes arose on the Sikkim and

EARLY TRAVELS.

Nepāl frontier, which, according to the terms of the treaty, were referred to the Governor-General. Accordingly, in 1828 General (then Captain) Lloyd was deputed to effect a settlement. In company with Mr. J. W. Grant, the Commercial Resident at Mālda, he penetrated the hills, which were still a *terra incognita* to the British, as far as Rinchinpong, and during this journey was attracted by the position of Darjeeling.

From a report* dated the 18th June 1829, in which he claims to have been the only European who ever visited the place, we learn that Lloyd visited "the old Goorka station called Dorjeling" for six days in February 1829, and "was immediately struck with its being well adapted for the purpose of a sanitarium." He seems to have been a little apprehensive of the rigours of winter, but, he added, "should the climate prove too cold, Ging, which is below it, and to which there is very easy access, would remedy the evil." On all grounds, he strongly urged the importance of securing possession of the place, and, in particular, pointed out its advantages as a centre which would engross all the trade of the country, and as a position of great strategical importance, commanding the entrance into Nepāl and Bhutān. Darjeeling itself, though formerly occupied by a large village and the residence of one of the principal Kāzis, was deserted, and the country round it was sparsely inhabited; but, he said, "if this part of the hills was resumed by us, or ceded, the Chief and people who have emigrated would instantly return, and, as he is very tyrannical, I don't suppose a single Lepcha would remain subject to the Sikkim Raja. I think it probable that they might also, in the space of a few more years, prefer the Christian to the Lama religion." At the same time, Mr. Grant also impressed on the Governor-General, Lord William Bentinck, the numerous advantages promised by the establishment of a sanitarium at Darjeeling, and strongly advocated its occupation for military purposes, as the key of a pass into the Nepāl territory.

Cession of Darjeeling. These representations were not neglected by Lord William Bentinck, who promptly deputed Captain Herbert, then Deputy Surveyor-General, to examine the country in company with Mr. Grant, observing that to the extreme earnestness of the latter in commending Darjeeling, that place would be mainly indebted for any importance which it might eventually obtain. The reports of these two gentlemen conclusively proved the feasibility of establishing a sanitarium at Darjeeling; and the Court of Directors approved the project, on the ground that it might prove a valuable depôt for the temporary reception of European recruits, and

* Dorjeling, by H. V. Bayley, 1838.

even a permanent cantonment for a European regiment. General Lloyd was, accordingly, directed to open negotiations with the Rājā of Sikkim, on the first convenient occasion, for the cession of Darjeeling in return for an equivalent in money or land This opportunity occurred in 1834-35, when some Lepchā refugees in Nepāl having made an inroad into the Sikkim Tarai, General Lloyd was deputed to enquire into the causes of the disturbance. The refugees were obliged to return to Nepāl, and the negotiations ended in the execution by the Rājā of Sikkim of a deed of grant on the 1st February, 1835.

This deed of grant, which is commendably short, runs as follows :—" The Governor-General having expressed his desire for the possession of the hill of Darjeeling on account of its cool climate, for the purpose of enabling the servants of his Government, suffering from sickness, to avail themselves of its advantages, I, the Sikkimputtee Rajah, out of friendship for the said Governor-General, hereby present Darjeeling to the East India Company, that is, all the land south of the Great Runjeet river, east of the Balasur, Kahail and Little Runjeet rivers, and west of the Rungno and Mahanuddi rivers." This was an unconditional cession of what was then a worthless uninhabited mountain; but in 1841 the Government granted the Rājā an allowance of Rs. 3,000 as compensation, and raised the grant to Rs. 6,000 in 1846.

EARLY SETTLE

The hill territory of Darjeeling having thus been ceded, General Lloyd and Dr. Chapman were sent in 1836 to explore the country, to ascertain the nature of the climate, and to investigate the capabilities of the place. Here they spent the winter of 1836 and part of 1837; and on receipt of their reports, it was finally decided to adopt Darjeeling as a sanitarium. General Lloyd was appointed Local Agent with power to deal with the applications for land which soon began to pour in from the residents of Calcutta, and the new settlement progressed rapidly. When General Lloyd and Dr. Chapman visited Darjeeling in 1836, all they found was some huts recently erected by the Rājā of Sikkim, in which they spent a night shivering with cold, without food or bedding. By 1840 a road had been made from Pankhābāri; there was a staging bungalow there and another at Mahaldiram; a hotel had been started at Kurseong and a second at Darjeeling; and at the latter place some 30 private houses had been erected, and nearly as many more locations had been taken up at Lebong.

Elsewhere, however, the country was still uncleared, the whole road from Pankhābāri running through virgin forest with an almost impenetrable growth of underwood broken only here and there by a Lepchā clearing, a mountain slip or a slope cleared by fire. The country was still practically uninhabited, and one of the most important problems of administration was to attract native settlers. About 10 years previously 1,200 able-bodied Lepchās, forming, according to Captain Herbert, two-thirds of the population of Sikkim, had been forced by the oppression of the Rājā to fly from Darjeeling and its neighbourhood, and to take refuge in Nepāl. What little cultivation there was, had been abandoned; the Sikkim Rājā had prohibited his subjects from going to Darjeeling and helping in establishing the new settlement; and various expedients were proposed to repopulate the country, *e.g.*, to invite the Lepchā refugees to return, to import labourers from the indigo concerns in Rangpur and Rāmgarh (*i.e.*, Gayā and Hazāribāgh), or to procure settlers from Nepāl and Bhutān.

EARLY ADMINISTRATION.

In 1839 Dr. Campbell, a member of the Indian Medical Service, who was British Resident in Nepāl, was transferred to Darjeeling as Superintendent. As such, he was in charge of the political relations with Sikkim and was entrusted with the civil, criminal and fiscal administration of the district, besides being Postmaster, Marriage Registrar and Administrator of the Station Funds, *i.e.*, the income from the lands in and about the station which had been leased for building purposes. All these duties were discharged by him single-handed, but Dr. Campbell devoted himself with rare energy to the task of developing the station, attracting immigrants to cultivate the mountain slopes, and stimulating trade and commerce. Every encouragement was given to settlers, who received grants of forest land; and the success which attended his efforts may be gauged by the fact that he was able to report that the population rose from not more than 100 souls in 1839 to about 10,000 in 1849, chiefly by immigration from the neighbouring States of Nepāl, Sikkim and Bhutān, in all of which slavery was prevalent.

"Whatever has been done here," wrote an inspecting officer* in 1852, "has been done by Dr. Campbell alone. He found Darjeeling an inaccessible tract of forest, with a very scanty population; by his exertions an excellent sanitarium has been established for troops and others; a Hill Corps has been established for the maintenance of order and improvement of communications; no less than 70 European houses have been built, with a bazar, jail, and buildings for the accommodation of the sick in the depôt; a revenue of Rs. 50,000 has been raised, and is collected

* Report on Darjeeling, by W. B. Jackson, C.S.—Selections from the Records of the Bengal Government, No. XVII.

punctually and without balance; a simple system of administration of justice has been introduced, well adapted to the character of the tribes with whom he had to deal; the system of forced labour formerly in use has been abolished, and labour with all other valuables has been left to find its own price in an open market; roads have been made; experimental cultivation of tea and coffee has been introduced, and various European fruits and grapes; and this has been effected at the same time that the various tribes of inhabitants have been conciliated, and their habits and prejudices treated with a caution and forbearance which will render further progress in the same direction an easy task. I may, in short, say of him that to him is the Government indebted for the formation of the district of Darjeeling."

RELATIONS WITH SIKKIM.

In the meantime our relations with Sikkim had been far from satisfactory. The Rājā, old and infirm, was a mere cipher in the hands of his minister Namguay, popularly known as the *Paglā Dīwān* or mad Prime Minister of Sikkim. Wrapt in religious contemplation, and abstracting himself from the consideration of earthly things, the Rājā allowed the Dīwān, a corrupt and ambitious official, to administer the State and enrich himself at its expense. The increasing importance of Darjeeling under free institutions was a source of early and constant jealousy and annoyance to the Dīwān, who was himself the monopolist of all trade in Sikkim; and it was shared in by the Lāmas and other notabilities, who lost their rights over slaves settling as British subjects in our territory. The plan pursued was to frighten our new subjects by spreading false reports and sending secret emissaries, by declaring that they should be delivered up as escaped slaves to their former masters, and by discouraging in every way the resort of the people of Sikkim to Darjeeling. Besides this, British subjects were constantly being kidnapped to be sold into slavery, and there were frequent denials of aid in capturing and surrendering criminals. "Every obstacle," wrote Sir Joseph Hooker, "was thrown in the way of a good understanding between Sikkim and the British Government. British subjects were rigorously excluded from Sikkim; every liberal offer for free trade and intercourse was rejected generally with insolence; merchandise was taxed, and notorious offenders, refugees from the British territories, were harboured; despatches were detained; and the vakeels, or Raja's representatives, were chosen for their insolence and incapacity. The conduct of the Diwan throughout was Indo-Chinese; assuming, insolent, aggressive, never perpetrating open violence, but by petty insults effectually preventing all good understanding.

He was met with neglect or forbearance on the part of the Calcutta Government and by patience and passive resistance at Darjeeling. Our inaction and long suffering were taken for weakness, and our concessions for timidity."

The climax was reached in November 1849, when Sir Joseph Hooker and Dr. Campbell were suddenly seized and made prisoners, while travelling in Sikkim with the permission both of the Rājā and of the British Government. The object of the Dīwān was to force Dr. Campbell to relinquish the claims for the surrender of criminals; to make him, while in durance, agree to the dictation of the Dīwān regarding the giving up of escaped slaves; and to detain him until these enforced conditions should be sanctioned by Government. This method of enforcing demands by capture and detention was common with the turbulent tribes east of Nepāl, but in this instance was aggravated by the violence and various indignities to which their captors subjected Dr. Campbell. Foiled, however, by his declaration that whatever concessions might be extorted would be repudiated by Government, and intimidated by the characteristic threat of the Governor-General, Lord Dalhousie, that the Rājā's head should answer for it, if a hair of the head of either prisoner were hurt, the Sikkimese eventually released Dr. Campbell and Sir Joseph Hooker on the 24th December 1849, a little more than six weeks after their seizure.

ANNEXATION OF THE TARAI.

In February 1850 an avenging force crossed the Great Rangīt into Sikkim; but after remaining on the northern bank of the river for some weeks, they were recalled without any further demonstration. The grant of Rs. 6,000 per annum which the Rājā had received since 1846 was withdrawn, and the Rājā was further punished by the annexation of the Sikkim Tarai, which he had originally received as a free gift from the British and which was the only lucrative or fertile estate he possessed. The seizure was quietly effected by four policemen taking possession of the treasury, which, it is said, contained exactly six rupees, and by announcing to the villagers the confiscation of the territory to the British Government. At the same time, there was annexed to it the portion of the Sikkim hills bounded by the Rammān on the north, the Great Rangīt and the Tīsta on the east, and by the Nepāl frontier on the west, a tract of country containing about 5,000 souls. The result was to confine the Rājā to the mountainous hinterland, and to cut off all access to the plains except through British territory. The change was welcomed by the inhabitants, for it only involved the payment of a small fixed tax in money to the treasury at Darjeeling, instead of a fluctuating one in kind, with service to the Rājā and liability to further annoyance from the Dīwān. The new territory was placed under the management of the Superintendent of Darjeeling, under whose administration it soon became a valuable asset, owing to the increase of the population and its suitability for tea. The whole country thus annexed covered an area of 640 square miles, and its annexation was an important measure; for it made the British boundary march with Nepāl on the west and with Bhutān on the east, while it connected Darjeeling on the south with the British districts of Purnea and Jalpaigurī. Previously the district had been an enclave in Sikkim territory, and to reach it the British had to pass through a country acknowledging the rule of a foreign, though dependent, potentate.

THE TREATY OF 1861.

For some years after this the relations between Sikkim and the British Goverment proceeded smoothly. The Dīwān, however, though ostensibly dismissed, soon worked his way into power again through his wife, an illegitimate daughter of the Rājā, and the former outrages were deliberately renewed. Constant raids were made upon our territory, property was plundered, our subjects were carried off and sold as slaves or detained in Sikkim, and no redress could be obtained. The Rājā of Sikkim, now an old man of nearly 80 years, had relinquished all cares of State and retired to Chumbi in Tibet, and the Government was entirely in the hands of the Chief Minister, Dīwān Namguay, the man who had seized Dr. Campbell and Dr. Hooker in November 1849, and who was the real author of the raids into our territory. After 6 months' negotiations, reparation was refused, and it was resolved to take possession of the portion of Sikkim lying to the north of the Rammān and to the west of the Great Rangīt, and to retain it till our subjects were restored, the offenders given up, and security obtained against a recurrence of similar offences. In November 1860 Dr. Campbell crossed the Rammān with a small force and advanced as far as Rinchinpong. But he had only 160 natives and a complement of English and non-commissioned officers, and, when attacked, he was forced to retreat for lack of ammunition and to fall back on Darjeeling. Subsequently Colonel Gawler, at the head of a force of 2,600 men, including 2 mountain howitzers and a detachment of artillery, with Sir Ashley Eden as Envoy and Special Commissioner, started from Darjeeling on the 1st February 1861 and reached Tumlong, the Sikkim capital, early in March 1861. The Dīwān fled, the British force dismantled the forts, the old Rājā abdicated in favour of his son, and, on the 28th March, Sir Ashley Eden effected a treaty with the new Rājā.

This treaty stipulated that full compensation should be made to those of our subjects who had either been kidnapped or pillaged by the Rājā's people; it provided for full indemnification for the losses sustained in Dr. Campbell's retreat; it guaranteed the opening out of the country to trade, and the removal of all restrictions on travellers and merchants; it fixed the maximum rate of transit duties to be levied on goods between British India and Tibet; it provided for the construction of roads, and the security of those who traversed them; and lastly, it contained provisions for the banishment of Dīwān Namguay, and for the future good conduct of the Sikkim Government. This treaty was of great importance to the interests of Darjeeling, as it finally put an end to the vexatious annoyances to which its inhabitants were exposed, and secured full freedom for commerce. During our occupation of Sikkim, a good road was constructed from Darjeeling to the Tīsta, while the Sikkim authorities undertook to complete the remaining portion between the Cholā Pass and Tibet; and the first step was thus taken to tap the Tibetan trade.

THE BHUTAN WAR.

Trouble soon after this arose with the adjoining State of Bhutān. The Bhutanese were constantly engaged in aggressions on the frontier, and there was a series of outrages in which property was plundered and destroyed, lives were taken, and many innocent persons were carried off into captivity. In the beginning of 1862 news came that the Bhutanese were making hostile preparations for the purpose of entering our territory, and an attack on Darjeeling was apprehended. Troops were at once hurried up from Dinapore, and confidence on the frontier was restored. This was followed in 1863 by the despatch of a special mission to Bhutān under Sir Ashley Eden, who was charged with proposals of a conciliatory character, but was also instructed to demand the restoration of plundered property. These pacific overtures were insolently rejected by the Government of Bhutān. Not only were restitution for the past and security for the future refused, but the British Envoy was openly insulted in *Darbār*, and compelled, as the only means of ensuring the safe return of the Mission, to sign, under protest, a document by which the Government of India was to renounce the Bhutān Duārs on the Assam frontier. Sir Ashley Eden, who had been treated with gross indignity, at last succeeded with difficulty in leaving Punākha during the night, and returned to Darjeeling in April 1864.

After further fruitless negotiations, the Government of India determined to annex the Bengal Duārs and so much of the hill territory, including the forts of Dālingkot, Pasaka and

Dīwāngiri, as might be necessary to prevent the hostile or predatory incursions of the Bhutanese into the Darjeeling district and the plains below. A military force of sufficient strength was accordingly despatched into Bhutān in the cold weather of 1864. The opposition offered by the Bhutanese was generally of the most contemptible kind. They were found to be efficient only in throwing up stockades and offering resistance from behind them; and in the plains they proved to be as despicable a foe as could well be conceived. Their fortresses were captured with the greatest ease, and the whole of the Duārs was completely occupied by the middle of January 1865. In November 1865 the treaty extorted from Sir Ashley Eden was given up, and a fresh treaty was executed under which the Bhutān Duārs, with the passes leading into the hills, were ceded to the British in return for an annual subsidy. The whole of the Bhotiā possessions in the plains thus became British, and a slip of British hill territory lying on the eastern bank of the Tīsta was interposed between Bhutān and Sikkim. In 1866 this tract, which now forms the Kālimpong police circle, was added to the district of Darjeeling. This was the last addition to the district, which thus acquired its present dimensions.

DARJEELING IN 1866.

The year 1866 may be taken as marking an epoch in the history of Darjeeling. Peace was established within its borders; and thenceforward began the march of progress and civilization. Rapid progress was now at last made in the development of the communications of the district, which the Sikkim expedition of 1860 and the Bhutanese war the year before had shewn to be vitally essential. Between 1839 and 1842, Lord Napier of Magdala, then a young lieutenant in the Royal Engineers, had been engaged in laying out the station of Darjeeling and in making a road through the virgin forest to the Tarai. This road may still be seen along the top of the Senchal ridge from Ghum to Kurseong, after which it descends somewhat abruptly to Pankhābāri. This road was, however, both too narrow and too steep for wheeled traffic, and it was recognized that, in the interests of Darjeeling and for the development of the frontier trade, a road broad enough for carts and with an easy gradient was absolutely necessary. The construction of the Darjeeling Cart Road, that wonderful example of engineering work which subsequently made the alignment of a railway so comparatively easy, was accordingly begun in 1861; and at the time the construction of a broad metalled road from the Ganges to Silīgurī was pushed on. By 1866 the latter had been completed at a cost of $14\frac{2}{3}$ lakhs of rupees, the road proving a work of great magnitude, owing to the nature of the heavily flooded country it crossed; the Cart Road from Darjeeling to Kurseong had also been opened to traffic, and the lower section was approaching completion; while another road was being driven through the malarial Tarai, in order to link the Cart Road up with the road to the Ganges and so establish uninterrupted traffic with the plains.

From this time dates the first attempt to make the hills the home of European education in India. For some years Bishop Cotton had been advocating the establishment of hill schools for Europeans, and his efforts were supported by Lord Canning who pointed out in a celebrated minute how the domiciled English and Eurasians would, if neglected, become profitless, unmanageable, and a glaring reproach to the Government, while, if properly cared for, they might become a source of strength to British rule and of usefulness to India. The main result of this movement was the establishment of several hill schools, the first of which was St. Paul's School, which was transferred in 1864 from Calcutta to Darjeeling. To this period, too, must be ascribed the establishment of the cinchona cultivation, which has now made it possible for the fever-stricken natives of Bengal to obtain a cheap and simple remedy at every post office. In 1864 the first large cinchona plantation was started at Rangbī in the Rangjo valley, which after some years of doubt and disappointment was successfully developed. This plantation is now the centre of the manufacture of the cheap quinine and cinchona febrifuge which is within the reach of the poorest peasant. The tea industry of which the foundations had been laid many years previously was now firmly established as a commercial enterprise. In 1841 Dr. Campbell had started the experimental growth of the tea plant with a few seeds grown in Kumaon from China stock. Twenty years later the original plants had grown to a gigantic size, one of them, it is said, being 50 feet in circumference and 20 feet high. It was recognized that the soil and climate of Darjeeling were favourable to its cultivation; Dr. Campbell's example was followed by others; large plantations were established and companies were formed, until in 1866 there were no less than 39 gardens with over 10,000 acres under tea and an outturn of nearly half a million pounds. This rapid extension of tea cultivation had, however, resulted in reckless deforestation, and it was recognized that it was necessary to provide for the careful conservation of the forests, for their protection from wanton or unscientific destruction, and for the proper development of this part of the resources of the country. The Superintendent of the Calcutta Botanic Gardens was appointed Conservator of Forests for Bengal; and special measures were taken for the protection of the forests about Darjeeling, where the wholesale felling of timber, which had followed the introduction of tea-planting, had resulted in wide-spread denudation. Accordingly, all unassessed lands in the old hill territory above 6,000 feet in height, and all such lands between the Tarai and an elevation of 3,000 feet, were made over to the newly-established Forest Department and were reserved for forest purposes. This was the beginning of forest conservancy in Bengal.

The two most important factors in the development of the district have been the choice of Darjeeling for a health resort and the subsequent planting of tea in the hills. To Grant and Lloyd, Darjeeling is indebted for the discovery of the possibilities of this bracing climate to reinvigorate the wearied workers of the plains ; to Lloyd for the cession of the hill territory, which, as his widow has proudly recorded in his epitaph, was due to his exertions and to his personal influence with the Rājā of Sikkim ; and to Campbell for the introduction of the tea plant and for his unceasing and successful efforts to develop the resources of the district. Soon after the cession of Darjeeling had been effected began the first making of roads under the skilled supervision of Lord Napier and the laying out of the station which has now grown into the town of Darjeeling; from the appointment of Dr. Campbell as Superintendent dates the first immigration of Nepalese from the west and of plainsmen from the south who flocked in to exploit the land under the even-handed justice administered by Campbell. Formerly, beyond a few Lepchās and Limbus with their little clearings in the forests, an occasional raid from Nepāl, or a stray visitor from the table-lands of Tibet, the Darjeeling Hills were practically uninhabited. But, as already shewn, these health-giving breezes, these lofty hills with their great forests and wonderful mountain scenery, appealed to the British pioneers ; a number of cottages soon began to spring up on the slopes around what is now the town of Darjeeling, roads were laid out, and inter-communication with the plains became possible. An early account of the difficulties of the journey to Darjeeling in the olden time tells how it took the best part of a fortnight by boat, *pālki* and pony, and at an expense of about Rs. 300, to do what the ordinary traveller now accomplishes in less than 20 hours and at about a tenth of that cost. These facts in themselves are good illustrations of the marvellous progress made in the last half century.

DEVELOPMENT OF THE DISTRICT.*

* I am indebted to Revd. R. Kilgour, B.D., for this sketch of the development of Darjeeling.

But it was not till the starting of the tea industry that the most rapid strides were made. With the planting of tea, with its enormous demand for labour, the clearing of forest, the opening out of land, and lastly the introduction of the machinery required for the manufacture, we begin an entirely new chapter in the history of Darjeeling. The history of the growth of this particular industry is given in a subsequent article, and what we are concerned with here is merely to note the predominating influence it has had in the economic and industrial development of the district. For the tea plantations needed labour, more immigrants were consequently attracted to the district, these immigrants had to be fed, clothed and housed, and all this stimulated trade. Further, the plantations needed machinery, and the hillmen had to be taught the mysteries of the steam engine and the marvels of European engineering; blacksmiths, carpenters and other workmen had to learn the higher branches of the callings followed by their ancestors, and from these simple hill people had to be made the engine-drivers necessary to keep the machinery going. For this they shew great aptitude, and it is remarkable what good engine-men they prove after a little training. Then, to get up the hills the stores necessary for the industry, and to take down to the sea the products of their toil, the tonga, the old bullock-cart, the pack-pony, and the human carrier bearing enormous loads on his back were found to be insufficient. The result was an agitation for a railway, an agitation which had its practical effect in the establishment of the Darjeeling-Himalayan Railway, which was opened up as far as Darjeeling in 1881. We have no hestitation in saying that had it not been for the staple industry of tea with its great traffic up and down the hills, and for the fact that the great hill cart road had already been made by the Public Works Department, Darjeeling would have had long to wait ere a railway would have overcome the obstacles interposed by nature and have faced the huge initial expense of driving a line along the steep hill-sides, beneath precipitous crags and over furious mountain torrents. The industry is still almost entirely in the hands of Europeans, as the difficulties of manufacture and the need of large capital have deterred the hill people from taking to planting tea on their own account.

Another important factor in advancing the general development of the district has been the preservation and development of the forests, which supply timber for its bridges and buildings, wood for its tea-boxes, and firewood and charcoal for its households. Instead of the great waste of timber, traces of which are visible in many parts of the hills, there are now extensive forests managed by an organised Forest Department. Instead of the old primitive methods of exploitation, the axe and saw are used for felling, the steam saw is just beginning to be employed for cutting up the timber, and in a few places wire tramways are used for transport. Here, again, the tea industry has played a leading part in helping on the development of the forests, for thousands of maunds of firewood and thousands of feet of planking are required every year for the factory boiler and tea-boxes respectively. Other economic products have been tried in the hills with varying success. Quinine, for example, was planted in large areas in several of the tea-gardens—one Company, in fact, is still known as the Darjeeling Tea and Cinchona Co.—and a few private planters opened up their land for cinchona. These plantations have gradually disappeared, as the cost of manufacture and transport was found to be prohibitive for private planters after the fall in the price of quinine; and only the Government cinchona plantation at Mangpu and its subordinate gardens, which produce both quinine and a number of alkaloid bye-products, are now at work. There, too, the introduction of up-to-date machinery and the presence of a staff of trained chemists and horticulturists have had a most important influence on the country-side, an example of which may be seen in the improved agricultural methods which are manifest in some of the villages near these plantations. The cultivation of coffee was started at about the same time as that of tea on one or two gardens, but did not prove a financial success. There are still a few gardens, *e.g.*, Barnesbeg and Glenburn, where a small number of bushes are still yielding berries which give quite drinkable coffee. Tobacco, again, which is indigenous in the hills, and which may be seen growing round many native homesteads, has never been successfully cultivated or manufactured on a large scale and developed by capital. One planter in the Tarai made an experiment with it, which for one year gave tokens of success, but he found that the damage done by the rains was far too great for him to get any profitable results. Experiments with rubber and camphor are at present being made, in the case of rubber with great prospects of success, but these experiments are only in the initial stage, and it is too early to make any forecast as to their eventual result.

Native cultivation is still practically at the same stage as it was when the British first occupied the country, except that the old, nomadic, wasteful system of cultivation practised by the Lepchās has been given up, and the use of the plough, which they have learnt from the immigrant Nepalese, has extended. Several attempts have been made to introduce better seeds, improved ploughs and other implements; and these attempts have met with some success, some remarkably fine Indian-corn having been grown in Kālimpong from American seed; but the hill people, though very receptive in some respects, are inherently conservative as regards agriculture, and with them the old is always better. Several planters have laid out good orchards, and have suggested to the hillmen that here was another source of profit, but the native, patient as he is in many ways, knows not the patience of waiting even to another generation for the fruits of his planting. A few natives have tried market gardening, for which there is an increasing demand, but with no great success. Orange groves, peach trees and pine-apples are, however, finding more favour at the lower elevations.

There has been a similar absence of improvement in the industries followed by the natives. Dr. Campbell once made an interesting experiment with the manufacture of paper and started a factory with the object of producing paper like the native-made paper of Nepāl. He introduced five Nepāli paper-makers, and for some years supplied strong writing and packing paper, but the competition of the cheap paper made in the plains was too strong, and the infant industry soon languished and died. Copper smelting of a wasteful character also used to be carried on to some extent, as the old Sikkim mint required a good deal of copper for producing the "dumpy pice" which are now valuable only to numismatists. Copper is, however, still used largely in the making of vessels for household use. Generally, it may be said that little improvement has taken place in the old handicrafts. The tools and implements in universal use are very much what they have been for centuries past; the old wooden plough and a little *kodāli*, the primitive spade of their forefathers, have been handed down to the present generation practically unchanged. In spite of the fact that the planters have shown how an improved tool produces improved work, the native impassively holds to his old ways. The women weave at the looms their grandmothers used, and, if they will only keep to the good stuffs and dyes of the old weavers, produce excellent work. A weaving school in Sikkim, and a few European looms and spinning wheels, introduced by missionaries throughout the district, have only affected the few at present. The ordinary Lepchā and Bhutiā woman and her Nepāli sister are quite content to work in the old ways, and spend months over a gaudy Lepchā *chādar* or a Nepāli cloth, which will certainly, at least in lasting wear, repay the long time spent on it. The advent of visitors, and specially of the cold-weather tourists, has produced among the Lepchās a crowd of collectors of butterflies, moths and orchids, a profession in which they are experts. Living the free life of nature, they are born naturalists, and to this they add a certain shrewdness, for they soon get to know the scientific value of their collections.

It is confessedly difficult for an Indian to learn any trade but that of his forefathers; but in Darjeeling among a Mongoloid people great strides have been made in teaching the Nepalese new handicrafts. We have already mentioned how the Nepalese lads on the tea gardens soon learn the mysteries of the engine-room and the different processes of tea manufacture. This receptivity is evident on a larger scale at the railway workshops at Tindhāria, where after a few months' training hill lads are found to pick up very quickly different methods of work; and at the electric-light works at Darjeeling they show a similar quickness to grasp the practical details of electrical engineering. But most of the trades are just as they were. The advent of the European has meant much building; yet Darjeeling has not produced any native architect and hardly any but ordinary workmen. Though Kātmāndu produces lovely carved wood-work, the hillman here refuses to learn carpentering. He will become a mason, but his hammer and chisel are much as those of his fathers were, and he leaves to Chinamen and the immigrants from the plains the higher branches of the art of building. Still there are some employments which he still gladly take up. The lower classes have learned to be house-servants, the Lepchās being especially faithful and useful. The more educated seek posts as clerks, headmen and assistants on tea gardens, as guards on the railway, etc., and, like the Bengalis, all are anxious for Government service. A few have learnt the art of printing, and there are presses in Darjeeling owned, managed, and worked entirely by Nepalese. The medical profession has also at last found its votaries among the hill people, who for too long despised all medical relief, but now come in increasing numbers to the planters and hospitals for it. As shop-keepers, however, they have been ousted by the sharp plainsmen, and there are only a few Newārs following their hereditary occupation as shop-keepers.

In the development thus rapidly sketched the most important elements have been the local tea industry, with its varied needs and the advent of the European with his many wants; and in the accomplishment of the progress which has been made we must give the foremost place to education, both vernacular and English. Darjeeling has also been fortunate in having several officials, many tea-planters, and a succession of missionaries who have devoted themselves to advancing not only the moral and spiritual but also the material welfare of the people. Many have taught with their own hands, more have shown by their example and active influence the uplifting power of a progressive civilization.

CHAPTER III.

THE PEOPLE.

GROWTH OF THE POPULATION.

THE growth of the Darjeeling district affords the most remarkable instance of the expansion of population of which there is any record in Bengal. When the British first acquired the hill territory in 1835, it was almost entirely under forest, and what small population it had, had been driven out by the oppression of the petty ruler whom they replaced. It was, in fact, estimated that the whole of this tract, comprising 138 square miles, contained only 100 souls. This state of affairs was soon altered by Dr. Campbell, the first Superintendent, whose object it was to inspire the aboriginal people of the hills with confidence in British rule, to induce the neighbouring tribes to settle in the territory, and to render Darjeeling the commercial centre of the hills. In this he was eminently successful, and by the year 1850 he was able to report that the number of inhabitants had risen to 10,000. Sir Joseph Hooker, indeed, compared the progress of Darjeeling to that of an Australian colony, both as regards the rapid extension of buildings and the accession of native families from the surrounding countries. When in 1869 a rough census was taken of the inhabitants in this tract, it was found that they aggregated over 22,000 persons.

The first regular census of the district as now constituted was carried out in the cold weather of 1871-72, and the result gave a total population of 94,712 persons, the average density of the population being 81 per square mile. When the next census was taken in 1881, the population had increased to 155,179, or by more than 63 per cent.; but there can be no doubt that a large portion of this increase was due solely to the incompleteness and inaccuracy of the first census. However that may be, the expansion of the population was almost equally remarkable during the next 10 years, at the end of which, *i.e.*, in 1891, it amounted to 223,314, showing an increase of 43·4 per cent.—and this in spite of the fact that large numbers, filled with fears of the intentions of Government, as wild as they were vague, fled across the borders to Nepāl rather than face the enumeration. Progress was far slower during the last decade, the increase amounting to 11·5 per cent., and the population, as returned at the census of 1901, being 249,117. In other words, the number of inhabitants is now more than 2½ times as great as it was 30 years ago.

Influence of the tea gardens. This phenomenal growth of the population since 1872 is due to two main causes, the development of the tea industry and the influx of settlers to exploit the waste lands of the district. How great the effect of the growth of the tea industry has been may be judged from the few statistics given below. In 1872 the number of tea estates in Darjeeling was only 74 with an area of 14,000 acres planted with tea. In 1881, the number of these estates had risen to 153, and the acreage under tea to 30,000, the number of estates having thus increased by 200 per cent. and the acreage being more than doubled. Ten years later there were 177 gardens, and 45,000 acres were actually under tea. These figures sufficiently denote the prodigious development of operations whose characteristic is the employment of an enormous quantity of manual labour. To this rapidly increasing demand the local supply of labour was quite unequal; the result was an unexampled immigration; and at the census of 1891 it was found that no less than 88,000 persons resident in the district were born in Nepāl. During the decade ending in 1901 the tea industry passed through a period of depression, and its expansion was checked, the acreage under tea amounting in the year 1901 to 51,000 acres and the number of estates to 170. Even so, however, the census of that year showed that the tea-garden coolies and their children accounted for more than two-thirds of the total population.

Agricultural settlers. As regards the influx of agriculturists, its extent will be apparent from the exceptional growth of the population of the Kālimpong tract to the east of the Tīsta. This tract extends over 401 square miles, of which 213 square miles are covered by forest, while 10 square miles are occupied by tea gardens. The remainder, or 178 square miles, has been reserved for native cultivation. It was annexed from Bhutān in 1865, and was then estimated to have a population of 3,530 souls. This number increased in 1881 to 12,683, and to 26,631 in 1891 or by 110·0 per cent. in the decade. In 1901 the population had grown to 41,511 or by 55·9 per cent. This rapid expansion is due entirely to the immigration of agriculturists. When the tract was first taken over, it was composed almost entirely of forest and hill-top inhabited by the aborigines of the country. It was resolved to keep a certain portion as a reserve for native cultivators, who flocked in to this land of promise; so that, with the exception of the forest reserves and some tea grown along the slopes of hills, the forest has been largely cleared away and the land brought under cultivation. The bulk of the population is now Nepalese, the continued immigration from Nepāl being due to the pressure of the population on the land in the eastern portion of that principality. There is no attempt at forest conservancy there, and the land is overcultivated, so that the Nepalese are glad to come to the undeveloped country in Darjeeling. Besides this, there is a considerable influx of Nepalese coolies from tea gardens, who come with large savings and buy up good lands or clear holdings from jungle; and of Lepchās, who leave the more thickly populated parts of the district. Whether the volume of immigration will be so great in the future is very doubtful. The best lands have been taken up, those now being developed consist of the poorer and more remote lands, which have been the last to attract settlers, though they are largely taken up by those who can get no good land elsewhere, and there are now only 30 square miles of reclaimable jungle left.

European population.

In concluding this sketch of the growth of the population, reference may be made to the increase in the number of Europeans which has taken place. In the year 1872 the district did not contain more than 419 Europeans and 32 Eurasians; whereas the number of the former has risen to 1,309 and of the latter to 329. This result is due solely to the development of the tea industry, in which the supervising staff is almost entirely European, and to the establishment of the town of Darjeeling as a sanitarium. In some ways, it is surprising that the number is not greater. At one time, optimistic hopes were entertained that a large European colony would be established in the district. Brian Houghton Hodgson looked forward not only to the rearing of subtropical products under European supervision, but also to agricultural settlements by the British race; and he hopefully pointed out that with "the backing of fifty to one hundred thousand loyal hearts and stalwart bodies of Saxon mould, our empire in India might safely defy the world in arms against it." This expectation has not been fulfilled, nor is it likely that the European farmer will ever be able to compete with the cheap labour of the hillmen in rearing country crops. Similar failure met the early attempts of the Moravian missionaries to maintain themselves by industrial labour, and in the end a few of them turned tea-planters and secured some of the best land in the district.

Census of 1901.

As already stated, the result of the census of 1901 was to disclose a total population of 249,117, the number of inhabitants as a whole having increased by 11·5 per cent. since 1891. Every part of the district showed an increase except the Tarai, which is notoriously unhealthy. It was greatest in Kālimpong, where the waste land is rapidly being brought under cultivation by new settlers. The remainder of the head-quarters subdivision also showed a fair growth, the tea gardens alone adding 5,000 or more than 12 per cent. to their population. In the hilly portion of the Kurseong subdivision there was a slight increase; the tea garden population was slightly reduced, but this loss was more than counterbalanced by the access of new settlers for ordinary cultivation, and to some extent by the development of the town of Kurseong. In the Tarai there was a small decrease, partly due to fortuitous causes, viz., to the fact that a large number of temporary immigrants employed in road-making were enumerated at the previous census. Even so, however, the decrease would have been far greater, had it not been for immigration; and it is a significant fact that while the settled cultivators on the Government estates decreased by nearly 5 per cent., there was an addition of 11 per cent. to the population of the tea gardens, which depend so much on imported labour. This tract is infested by malaria, and the mortality is exceptionally high, so that the rate of expansion has always been less rapid than in the hills. In the nineteen years ending in 1891 the growth in the former was 222 per cent., and in the latter it was only 52 per cent.; and, as we have already seen, the Tarai was the only decadent portion of the district in 1901.

GENERAL CHARACTERISTICS.

Density of population.

The district, as a whole, is very thinly peopled, there being only 214 persons to the square mile. It must, however, be remembered that no less than 445 square miles, or 38 per cent. of the entire area, are occupied by reserved forests, and, if these be excluded, the density will rise to 346 persons per square mile. The pressure of the population on the soil is greatest in the Tarai, where there are 279 persons to the square mile; it gradually diminishes towards the north, and in the head-quarters subdivision there are only 184 persons to the square mile. The most sparsely populated tract is Kālimpong, which had in 1891 only 65 persons to the square mile. Here, in spite of the fact that the population increased by more than 55 per cent. in the succeeding decade, the density is still only 101 to the square mile.

Migration.

From the preceding account it will be observed that Darjeeling owes a large proportion of its population to the advent of immigrants Barely half of the inhabitants have, in fact, been born in the district. In the Tarai it receives numerous settlers from the adjoining districts of Purnea and Jalpaigurī, who engage in cultivation, and a large number of coolies from Chotā Nāgpur and the Santāl Parganas, who are attracted by the wages given in the tea gardens. But the great bulk of the immigrants come from Nepāl, chiefly as labourers in the tea gardens; and in the hills the population is mainly Nepalese. The early immigrants are, however, dying out, their place being taken by their children born in Darjeeling, while the flow of fresh immigrants is growing less. Most of the latter settle in the district, but some only come for a short time, and then return to Nepāl with their savings. As a result of this immigration, there is a large preponderance of males, there being only 87 women to every 100 men in the district

The volume of emigration from the district is very small. A certain number leave the district to serve in the Military Police battalions in Burma and elsewhere, but the great majority of emigrants consist of cultivators who throw up their holdings and cross the border to Sikkim or Bhutān, where the forests are abundant and land is to be had for the asking.

Towns and villages.

There are only two towns in the district, Darjeeling and Kurseong, which between them account for only 21,393 persons or nearly 8 per cent. of the population. The population of the former, according to the census of 1901, is 16,924 during the cold weather, but at a special census taken in September 1900 it was found to be 23,852 or nearly 50 per cent. more than in the cold weather. This difference is due to the fact that in the hot weather and rains it is the temporary head-quarters of the Bengal Government, and is largely resorted to by Europeans whose permanent residence is in the plains. Kurseong has a population of only 4,469 and is far from being progressive, the number of inhabitants being only 436 more than in 1881. Like Darjeeling, it is a hill station, situated at a lower elevation, but it does not enjoy the same reputation as a sanitarium, though it is admirably adapted for the purpose. Till recently it has shewn but little signs of development. This want of progress must be ascribed, to some extent, to an absence of enterprise among its inhabitants, but more largely to the want of a colonizing spirit among the Europeans of Bengal. To the same cause must be attributed the fact that Hope Town has never developed in spite of the central portion it occupies. It was at one time let out for building sites, and a church was erected, in the hope that it would develop into a small hill station, but the scheme proved abortive, and the land is now, for the greater part, under tea.

The only other places of any importance are Silīgurī and Kālimpong, the former being a village of some 784 souls in the Tarai, and the latter having a population over 1,000 inhabitants. The former is a swampy malarious village close to the foot of the hills, but Kālimpong is charmingly situated at a height of nearly 4,000 feet with a delightful winter climate and a rainfall far smaller than that of either Darjeeling or Kurseong. Part of the land has been set aside for European settlers, but it has failed to attract the Anglo-Indian, in spite of all its advantages, and the European population is confined to the members of the Church of Scotland Mission and the inmates of the Colonial Homes.

The remainder of the population is scattered along the hill sides and over the level flats of the Tarai, and consists of native cultivators tilling their own small fields or labourers employed in the tea gardens. Villages, in the proper sense of the term, are almost unknown. In the hills with the exception of the cooly lines on the various tea plantations, each homestead stands in its own land near the patches of cleared cultivation, though occasionally five or six houses are grouped together; and there is no corporate village life such as there is in the plains. Even in the Tarai the social unit is the farm, not, as elsewhere in India, the village. In each group of huts the principal buildings are occupied by a substantial cultivator, often the representative of the original reclaimer of the soil, the others by his labourers or tenants. He keeps in a store-house elevated on piles a large stock of rice, the resource of the whole hamlet. From it he furnishes with seed those entitled to it, and he frequently makes subsistence loans. He has to keep the whole body together, a task by no means easy, as the other members of the small community take but little interest in the matter, and are always ready to emigrate. There are good markets in the Tarai, but no villages. That ancient form of organization, which elsewhere dominates rural life, is here unknown; and its absence makes us feel that, when we enter the Tarai, we have passed the former limits of India, and are entering upon those of the Mongolian race.

Races of Darjeeling. The population of Darjeeling is exceedingly heterogeneous. The majority of the people in the hills are of Mongolian origin, belonging chiefly to various Nepalese castes, but also including a large number of Lepchās, Bhotiās and Tibetans. Together with these hillmen are found the denizens of the plains, who have been attracted to the hills by the prospect of easily acquired wealth, the Madhesias held in great contempt by the stalwart Nepalese. Among them are Mārwāri merchants, the Jews of the

Himâlayas, Bengali clerks, Hindustâni mechanics, Punjâbi traders and even Chinese carpenters. In the Tarai the mixture of races is equally great. Here the aboriginal Koches, or Râjbansis as they prefer to call themselves, are most numerous; numbering 29,460; but no less than 52 per cent. of the inhabitants were born elsewhere, and the Mundâs and Oraons from Chotâ Nāgpur and the Santâls from the uplands of the Santâl Parganas have a strength of nearly 14,000. Darjeeling has, in fact, been described as a "Babel of tribes and nations," and the extraordinary variety of races may be realized from the fact that at one end of the scale of civilization stands the European, and, at the other, remnants of races who express "agriculture" by the term "felling" or "clearing the forest," who have no term for "village," for "horse," for "plough," for "money" of any kind, and whose language is marked by an absence of any term for nearly every operation of the intellect of will, whether virtuous or vicious, and for almost every abstract idea, whether material or immaterial.*

The Nepalese.

The dominant race in Darjeeling is the Nepalese, which with a strength of 134,000 accounts for more than half the population. The national characteristics of the Nepalese have been so frequently described that only a brief mention of them is required. They are a capable, cheerful and alert people, and are essentially a virile race. Though quick-tempered and keen to resent an injustice, they are remarkably willing and loyal, if treated with consideration. Born cultivators, resourceful and hard-working, they have taught the Lepchâ aborigines much in the way of agriculture, and, with their greater thrift and industry, they have managed to supplant the latter and secure most of the best agricultural land in the district. Much of this they buy up with their savings, but often they merely annex it by quiet encroachment, presuming on the apathy and timidity of their Lepchâ neighbours. On the whole, however, they are remarkably law-abiding, a trait which is characteristically expressed by one of their commonest proverbs—"There is no answer to an order, there is no medicine for death." Their general character has been happily described by Colonel Waddell, who writes: "Though small in stature, these Nepalese have big hearts; and in many ways resemble the bright, joyous temperament of the Japanese, though lacking altogether the refinement of the latter. Naturally vigorous, excitable and aggressive, they are very law-abiding, driven as they have been to obedience by the draconic punishments

---

* Hodgson's Miscellaneous Essays, vol. I, p. 103.

of their Gurkha rulers. In appearance, the various tribes vary considerably, in proportion to the extent of their admixture with Aryan blood. Scratch a Russian it is said, and you'll find the Tartar; but the Nepalese, even with their thin veneer of Hinduism, do not require this operation to reveal their Tartar character. The features of the great majority are markedly Mongolian, with oblique eyes, and little or no moustache. They are generally undersized, but tough and wiry as whip-chord, and so full of energy that it is quite common to see old people scampering nimbly up and down hill in preference to walking*."

Nepalese castes. The various Nepalese castes are well represented in the district, but the most numerous are the Khambus and Murmīs.

Khambus. The Khambus, who number 33,000, are a caste living on the southern spurs of the Himālayas, whose home is the eastern portion of Nepāl between the Sankos river and the Singalīlā range and Mechī river. They claim to be Jimdârs, one of the fighting tribes of Nepāl and bear the Kirānti title of Rai. Their religion is partly Lāmaist Buddhism and partly Hinduism.

Murmīs. The Murmīs, who number 25,400, are a Mongolian or semi-Mongolian caste who claim to be among the earliest settlers of Nepāl. They are probably descended from a Tibetan stock which has been modified by intermixture with Nepalese races. They are also known as Tamāng Bhotiās and bear the title of Lāma. The bulk of them are cultivators and regard agriculture as their original and fitting occupation. Many of them, however, serve in the police and some in the army, and a very large proportion in this district are employed as labourers on the tea gardens. They are gradually adopting the Nepalese form of Hinduism, and Buddhist usages are on the decline, though Lāmas still serve as priests at their weddings, and flags stamped with the Buddhist formula, the sacred *Om mani padme Hum*, may be seen flying from their homesteads.

Limbus. The next most numerous Nepalese caste is the Limbu (14,300), who bear the title Subha and whose original home is also in the east of Nepāl. Their flat features, slightly oblique eyes, yellow complexion and beardlessness give some grounds for supposing that they are the descendants of early Tibetan settlers in Nepāl; but they have now intermarried largely with Lepchās. Though considering themselves a military race—and they offered a most gallant resistance to the invading Gurkhas—they do not rank among the regular fighting tribes of Nepāl; they are now

* Among the Himalayas.

chiefly carriers and porters, but also engage in agriculture, grazing and trade.

Three other numerous castes are the Khas or Chettri (11,600), the Mangars (11,900) and the Gurungs (8,700), these being the three dominant tribes of Nepāl which overthrew the Newār dynasty in 1769. The Khas, who have a large admixture of Aryan blood, belong to the proud military order of Nepāl, which has adopted the title Chettri as a caste name. They and the Nepāli Brāhmans, who number nearly 5,000, make careful and successful cultivators in this district. The chief occupations of the Mangars are agriculture, trade and soldiering, but no means of earning a livelihood comes amiss to them, provided it does not involve doing things which are properly the business of the lower castes. The Gurungs are a nomadic pastoral race, who still obtain their subsistence by rearing and grazing cattle. Another Nepalese caste fairly numerous in Darjeeling is that of the Newārs (5,880) originally semi-aborigines who were the ruling race of Nepāl until ousted by the Gurkhas. Newār is, in fact, the name of a nationality rather than a caste, Nepāl and Newār being two forms of the same word; and it means simply a member of the community which inhabited Nepāl proper prior to the Gurkha conquest. The caste bear the title of Pradhān. They are now traders and artisans, agriculturists and domestic servants, and some still adhere to their old religion Buddhism.

Other castes.

Among other Nepāli castes may be mentioned the Yākhās (1,143), an agricultural caste calling themselves Dīwān, who come from the same tract of country as the Khambus and Limbus, and are considered by some to be merely a sept of Jimdārs, and the Sunuwārs, a cultivating tribe who were originally hunters. Among the lower castes the most numerous are the Kāmis (9,800) who are blacksmiths and goldsmiths, the Damai (4,600) or tailor caste, the Sarki (1,800), who work in leather, and the Gharti (3,450) who are the descendants of manumitted slaves. These four castes are so low that in Nepāl they may not enter the courtyards of temples, and have to leave the road on the approach of a member of the higher castes and to call out to give warning of their approach.

The caste system is however by no means strict among the Nepalese domiciled in Darjeeling, where the Brāhman may be found working as a cultivator, a labourer or even as a *sais*. There is an extraordinary laxity in ceremonial observance; they will eat and drink things which are an abomination to the orthodox Hindu of the plains, and many of them are great flesh-eaters, relishing even beef and pork. The contrast in this respect between the Nepalese and the plainsmen has been forcibly pointed out by Brian Houghton Hodgson, who wrote: "These highland soldiers, who despatch their meal in half an hour, and satisfy the ceremonial law by merely washing their hands and face and taking off their turbans before cooking, laugh at the pharisaical rigour of the sepoys, who must bathe from head to foot, and make puja ere they can begin to dress their dinners, must eat nearly naked in the coldest weather, and cannot be in marching trim again in less than three hours."

The Lepchas.

The Lepchās are the aboriginal inhabitants of the country, who call themselves Rong, *i.e.*, the squatters, and their country the land of caves. The word Lepchā, or as it should be spelt Lapcha or Lapche, means the people of vile speech and was a contemptuous appellation given to them by the Nepalese. Formerly they possessed all the hill country of Darjeeling and Sikkim, but about 250 years ago the Tibetans invaded their country and drove them into the lower valleys and gorges; and in 1706 the tract east of the Tīsta, which is now included in Kālimpong, was conquered and taken from them by the Bhutanese. For generations past a conquered race, they are a timid people, peaceful and no brawlers, disliking fixed employment and never so happy as when they are in their native woods. They are born naturalists, and have separate names for nearly every bird, plant, orchid and butterfly. With the introduction of settled cultivation and the reservation of the forests, they have had to give up their old nomadic cultivation, and have lost much of their jungle craft; but they have learnt in its place how to make terraces for rice fields and the methods of agriculture practised by the Nepalese. A kindly, placid, somewhat indolent people, they are being largely supplanted by the pushing, self-assertive Nepalese, and in consequence are losing the best land in the district. In spite, however, of their unenterprising nature, they often make excellent servants; and their want of vigour has been much exaggerated, as will be apparent from the fact that some of them have gone far afield and done excellent entomological work in Burma, the Andamans and Nicobars, Sumatra, Borneo, and the Malay Archipelago, in the Celebes and New Guinea, where one of them was killed by the savages, and in Central Africa. Here, the European in whose service they were, died far away from civilization, but the Lepchā collectors contrived to get back to Darjeeling with the help of the long arm of British authority. Such a feat goes far to show that they are not such a resourceless people as is often supposed.

It is often said that they are disappearing with the forests they love, and that they are a dying race. This, however, does not appear to be the case, so far as any conclusions can be based on the census statistics of the last 30 years. According to the census of 1872, there were only under 4,000 Lepchās in the district, whereas at the last census of 1901 their numbers were found to be nearly 10,000. They are not so prolific as the Nepalese—their numbers are indeed practically the same as in 1891—but they perpetuate their families by the practice of adoption. At the same time, however, the race has become somewhat impure by intermarriage with other races, especially with the Limbus and Sikkim Bhotiās, who stand higher in the social scale, and in this way the Lepchās are contributing to their self-effacement. The tendency is for them to leave the unsuitable environment of Darjeeling and other parts of the district, and to settle in Kālimpong, where they can get land on low rents and where is still much forest left. Many, however, emigrate to Bhutān, where still more abundant forests, untrammelled by restrictions, enable them to follow their ancient but wasteful system of *jhmūing* or shifting cultivation, which consists of burning down a fresh patch of jungle each successive year, and raising crops in the ashes. It should be remembered, moreover, that they have never been very numerous in this district. When we first acquired the hill territory of Darjeeling, there were practically no Lepchās in the land, and a contemporary writer says that the oppression of the Rājā had forced 1,200 able-bodied Lepchās, who, he says, formed two-thirds of the population of Sikkim, to fly from Darjeeling and its neighbourhood. Even without accepting this estimate as accurate, it serves to show how sparse the population of Lepchās was at that time, and to indicate that the Lepchās have at any rate not decreased under British rule, largely no doubt as a result of the establishment of a reserve for them in Kālimpong.

The Bhotiās of Darjeeling number 9,300 and consist of four classes, the Sikkimese Bhotiās, who are a mixed race of Tibetans and Lepchās, being the descendants of Tibetans who settled in Sikkim a few centuries ago and intermarried with Lepchās; Sharpā Bhotiās, who come from the east of Nepāl, the word *shar* merely meaning east; the Drukpā or Dharmā Bhotiās, whose home is in Bhutān; and the Tibetan Bhotiās from Tibet. The great majority of the first class, which numbers 1,550, are found in and around Darjeeling, of the second, which numbers 3,450, to the west of the district, while the third class, which numbers 2,350, is practically confined to Kālimpong, where they are the descendants of the Bhutanese who were settled on the land at the time of the annexation in 1865. There they have been reinforced by the immigration of Tibetans from the Chumbi valley and

The Bhotiās.

its neighbourhood, who have been attracted by a more fertile country and a more regular form of government than that of Tibet. The fourth class, the Tibetan Bhotiās, who number 1,700 souls, are pure Tibetans, who have immigrated from the tableland of Tibet.

The word Bhotiā means properly an inhabitant of Bhot or Tibet, and is synonymous with Tibetan. The native name of Tibet is Bod, and the Sanskrit form of this word was Bhot. The Sanskritic-speaking races of India have accordingly called the inhabitants of this region Bhotiās. The country of Bhutān was so called by the Bengalis in the belief that it was the end of Bhot (Bhotānta), and the natives of Bhutān, as well as of Tibet. are indiscriminately called Bhotiās. The English word Tibet, appears to be derived from the Mongolian Thübot, which is the Mongolian name for the northern portion of the Tibetan plateau.

As a race, the Bhotiās have been described as rude, turbulent and quarrelsome, but this seems an unfair estimate of their character. On the whole, they are a merry, cheerful people, quick to enjoy a joke, and most willing workers, not so pushing as the Nepalese nor so law-abiding as the Lepchā. Powerfully built and of great natural strength, they are capable of carrying the heaviest burdens—there is a story current that in the days before the railway a single Bhotiā carried a grand piano up the hills to Darjeeling, 50 miles distant and 7,000 feet in elevation;—but their natural love of display and an inordinate love of gambling soon dissipate the sums which they can easily earn by labour.

The Rājbansis.

The Rājbansis, or Koches as they should strictly be called, with a numerical strength of nearly 30,000, form the predominant race in the Tarai. They are the remnants of an aboriginal race, the Koches, who formerly possessed a powerful kingdom at the foot of the hills. They gradually became Hinduized by contact with their neighbours in the plains, and discarded the name of Koch for the magniloquent title Rājbansi, *i.e.*, the scion of kings. According to one view, they are a Mongoloid race that entered Bengal from the east by way of the Brahmaputra valley. Others consider it more likely that they are descended from a Dravidian stock which probably occupied the valley of the Ganges at the time of the Aryan advance into Bengal, that driven forward by this incursion into the swamps and forests of the Tarai, they were brought into touch with the Mongoloid races of the Lower Himālayas, and the type became affected by intermixture with them. A former Deputy Commissioner of Darjeeling, Mr. W. B. Oldham, described them as "the most conspicuously Dravidian race in Bengal. Their men are tall and robust, and neither in feature nor dress strikingly Tamulian; but at a market or village gathering thronged by their women one could imagine oneself transported to Kanara or Tamil land. The Rājbansi women, whose cast of feature is singularly homely and rough-hewn, leave the head uncovered and wear a dress in which blue or purple invariably predominates, reaching only to the knee and bound over the bosom, leaving both shoulders bare in a fashion not seen among other people in Bengal; while the ornaments of the head and limbs recall those worn in Southern India. Their language, however, is only Bengali, and they have no separate dialect or patois. That their adhesion to Hinduism is comparatively recent is shewn by their own customs as regards burial, food and marriage, as well as by the existence of the Dhimāl, who might be defined as a non-Hinduized Koch or Rājbansi among and beside them." There seems, however, to be no doubt that the true Koches were a Mongoloid race, and we find that in Jalpaigurī, Cooch Behār and Goālpāra, the persons now known as Rājbansi are either pure Koches who, though dark, have a distinctly Mongoloid physiognomy, or else a mixed breed, in which the Mongoloid element usually preponderates.*

Darjeeling contains a polyglot population. In the hills Nepāli Hindī and in the Tarai Bengali are spoken by at great majority of the inhabitants, but there are a great number of other languages current. Nearly half the people speak languages of the Tibeto-Burman family, of which no less than 19 different dialects were shewn in the census returns of 1901. Tibetan is spoken by the Bhotiās domiciled in the district, who, according to their nationality, speak the Tibetan or Bhotiā of Tibet, Sikkim or Bhutān; while the aboriginal Lepchās still to some extent speak their native Lepchā, or as they call it Rong-ring, a language which General Mainwaring believed to be the oldest in the world. In the hills Limbu is another fairly common dialect, and in the Tarai Dhimāl is still used by the remnants of the primitive race of Dhimāl aborigines. Among the Nepālis of Darjeeling Khambu (Jimdār), Murmī and Mangar, are the commonest tribal dialects, but Newār, Gurung, Sunuwār and Yākhā are also spoken. The plainsmen who have made their homes in Darjeeling speak Hindustāni, while in the Tarai tea gardens Oraon, Mundāri and Santāli are spoken by the coolies from the Chotā Nāgpur plateau. LANGUAGES.

Nearly one-fifth of the whole population speak Khas, *i.e.* Nepāli Hindī, or as it is sometimes generically called Pahāria or Parbatiya. This is a form of Hindī spoken by the

* *See* Gait's History of Assam, pp. 44-45.

Khas tribe of Nepāl, who obtained it from the numerous Brāhman and Rājput refugees that took refuge in the hills, when the ancient Hindu kingdoms were overthrown by the Muhammadan invaders. There they intermarried and gradually became fused with the original Mongoloid inhabitants, upon whom they imposed their religion and language. Since the overthrow of the Newār dynasty in Nepāl in the 18th century, Khas or Khaskura as it is also called (*kura* merely being an affix meaning speech) has gradually spread throughout Nepāl and beyond its borders. It is gradually ousting the various tribal dialects, and is now current as a *lingua franca* both in the principality of Nepāl and the polyglot district of Darjeeling.

RELIGIONS. At the census of 1901, over 187,000 persons or more than three-quarters of the population returned themselves as Hindus; and the total number of Buddhists, who are mostly Lepchās, Bhotiās and Murmīs, was 44,000. The Muhammadans formed less than 4 per cent. of the population, most of them being residents of the Tarai, where many of the aboriginal Koches have embraced Islām; while Animists accounted for 1·38 per cent. and Christians, with a strength of 4,467, for 1·8 per cent. of the total population.

Hinduism. Broadly speaking, the Hinduism professed in the district is nothing more than a thin veneer over animistic beliefs. Beneath this veneer the real popular religion can be seen in the worship paid to a host of spiritual beings whose attributes are ill-defined, but whose chief power is to cause evil to their votaries. The religion prevalent is in fact demonolatry, of which exorcism and bloody sacrifices are the most prominent feature. Throughout the hills there are signs of the prevailing fear of demons, such as "the little offering in the middle of the path to bar the progress of an evil spirit or the living sacrifice being offered to propitiate another, or the flattering rice image of a demon supposed to be causing sickness, or the burning of a rag before the door, over which the friends step when they return from burying a relative, to prevent any accompanying spirits from entering with them.*"

In the Tarai the same fear and worship of evil spirits prevails. The Rājbansi, whose greatest enemy is the deadly Tarai fever, has three chief deities whom he appeases by offerings of goats, ducks, etc., viz., Kālī Thākurānī, "the mother of sickness;" the Grām Devatā, "the goddess who prowls round village sites to cause illness among children," and Bishhaarī Thākurānī, "the source of all pains"—a trinity which would scarcely be selected by a happy and contented race Should

* On the Threshold of Three Closed Lands, p. 104, by The Revd. J. A. Graham, D.D.

drought last long, the Rājbansi women make two images of mud or cowdung, which are supposed to represent a god called Hudum Deo. This they carry away into the fields at night, and there strip themselves naked, and dance round the images, singing obscene songs, in the belief that this will cause rain to fall. The household god is represented by a round lump of clay made smooth by smearing it with cowdung. This is set up before a bamboo, and offerings are made to it of rice, which is afterwards eaten by the worshippers. If this duty is neglected, disease or some other calamity is sure to visit the family.

Buddhism. The form of Buddhism* prevalent is not of a much higher type. The craving for protection against malignant gods and demons causes the people to pin their faith on charms and amulets, and to erect tall prayer-flags, with strings of flaglets, which flutter from house-tops, bridges, passes and other places believed to be infested by evil spirits. "Prayers hang upon the people's lips. The prayers are chiefly directed to the devils, imploring them for freedom or release from their cruel inflictions, or they are plain naive requests for aid towards obtaining the good things of this life, the loaves and the fishes. At all spare times, day and night, the people ply their prayer-wheels, and tell their beads, and mutter the mystic six syllables—*Om mani padme Hum!* 'Om! the jewel in the lotus, Hum!' the sentence which gains them their great goal, the glorious heaven of eternal bliss."* This demonolatry has been aptly described by Doctor Graham in his book "On the Threshold of Three Closed Lands." "To the aboriginal Lepcha," he says, "the rites of religion are chiefly valuable in averting the anger or malice of an evil spirit, and all sickness is caused by such possession. The *Bongting* or sacrificial priest is the cunning expert who indicates the offended demon, and prescribes the proper sacrifice of cow or pig or goat or fowl needed to appease him. As a perpetual offering to ward off danger, each household keeps in one corner a little basket, containing rice and a small silver coin. Time and again it has happened that families who had sacrificed all they possessed—nay, had also drawn a bill on the future by putting into the votive basket a little rice tied up in a leaf as a pledge of a coming offering when they could afford it—had found all to be unavailing; and after they had come to an end of their resources, had called in the Christian catechist."

* The Buddhism of Tibet or Lamaism, by Lt.-Col. Waddell, I.M.S.

Animistic religion.

On the whole, Hinduism seems to be spreading at the expense of Buddhism. Cut off from his home, the Buddhist Nepāli soon adopts the religious beliefs of his neighbours, and, in the absence of a Buddhist priest, is fain to accept the ministrations of Brāhmans and to enrol himself, nominally at least in the ranks of Hinduism. This is all the easier owing to the fact that the popular religion both of the Hindus and the Buddhists is based on demonolatry and that there is no deep cleavage between the two sects. Even the Brāhman joins the other peasants, both Hindus and Buddhists, in making regular contributions to the Buddhist monasteries in order that the Lāmas may protect his crops from hail. How narrow a dividing line there is between the forms of the two religions in the hills may be seen from the practice of the Limbus.* The phlegmatic and utilitarian habit of mind which is characteristic of the Mongolian races comes out conspicuously in the nonchalant attitude of the Limbus towards religion. Where their surroundings are Hindu, they describe themselves as Saivas, and profess to worship, though with sparing and infrequent observance, the deities most favoured by the lax Hinduism of Nepāl. In a Buddhist neighbourhood the yoke of conformity is still more easy to bear : the Limbu has only to mutter the pious formula, *Om mani padme Hum*, and to pay respect and moderate tribute to the Lāmas, in order to be accepted as an average Buddhist. Beneath this veneer of conformity with whatever faith happens to have gained local acceptation, the vague shapes of their original Pantheon have survived. Whether concealed by the thin veil of Buddhistic or of Hindu usage, the primitive animistic belief persists. "The rude mind," it has been said, "with difficulty associates the idea of power and benignity. A shadowy conception of power that by much persuasion can be induced to refrain from inflicting harm, is the shape most easily taken by the Invisible in the minds of men, who have always been pressed close by primitive wants and to whom a life of hard toil has never been illuminated by any enthusiastic religious faith." The religion of the majority of the population is of this type. The current belief is that there are a number of malevolent spirits to whom the ills of life are due, and who exercise their malicious influence on the bodies and minds of men by means of demoniacal possession. Worship, therefore, consists of periodical propitiation of them in order to escape their attacks, and the cure for disease is not medicine but exorcism.

* *See* articles on Limbus and Mangars in Risley's Tribes and Castes of Bengal.

Progress of Christianity.

In 1829, when urging upon Lord William Bentinck the desirability of securing Darjeeling as the site of a hill sanitarium, General Lloyd hazarded the opinion that the people might, in the space of a few more years, prefer the Christian to the Lāma religion. These hopes were doomed to disappointment. The first attempt to introduce the doctrines of Christianity among the hill tribes was made in 1841 by Mr. Start, a Baptist, who brought out from Europe a small band of German missionaries, and began work among the Lepchās in Darjeeling and its neighbourhood, the head-quarters of the Mission being at Takvār. These Moravian missionaries endeavoured to maintain themselves by labour, but the scheme was soon found impracticable, as they were unable to compete with cheap native labour. Many of them took to the more lucrative employment of tea-planters, and their descendants are now among the oldest planters in the district.

The real beginning of missionary enterprise on an organized scale may practically be said to have been made in 1870 In that year the Church of Scotland started work among the aborigines of the district; one division consisted of a short-lived mission in the Tarai, two German missionaries being sent out to preach to the Meches and Rājbansis; in the other was Mr. Macfarlane of the same Church who came from Gayā and was the pioneer of Christianity among the hill tribes. His efforts were singularly successful, and he had the satisfaction before his death in 1887 of seeing over 600 native converts attached to the Mission. The number of native Christians has increased wonderfully since that time, and now amounts to nearly 3,000, the bulk of whom are Lepchās in the Kālimpong subdivision. There are now four divisions of the Mission :— (1) the Darjeeling Division, including Kurseong and the Tarai, from which have developed —(2) the Kālimpong or Guild Mission, which includes the Duārs and is so called because, while part of the whole Eastern Himalayan Mission of the Church of Scotland, it is supported by the Guilds of the Church, (3) the Universities Mission, which is supported by the Scottish Universities and works in Sikkim and at the Training School at Kālimpong, and (4) the Ladies Mission at Darjeeling and Kālimpong, which works among the girls and women of the hills.

The work is conducted in the Hindī, Nepāli, Lepchā and Tibetan languages. There are two noteworthy features in its organization, viz., that the affairs of each Christian community are managed by its own *panchāyat* or Presbytery, and that the branches established in the various villages are self-supporting, building their own churches and paying partially for their pastors. In the words of a former Lieutenant-Governor "the assistance given by this Mission to the work of civilization has been considerable. It has been the agent of Government in the spread of education. It has co-operated with the District Officer in keeping order in the village, and in putting down drunkenness, gambling and other vices. Turning to the more definitely religious side, the missionaries have been unusually successful in converting the simple tribes from their animistic or Buddhistic beliefs to the Christian faith".

Another Protestant mission is the Scandinavian Alliance Mission at Ghum, which works among the Bhotiās and Lepchās of Sikkim and Darjeeling. It was started in 1892 by some missionaries who came out from America with the intention of going to Tibet, but, finding that impossible, remained at Ghum and started mission work there. The last important mission is the Roman Catholic Mission. The portion of the district west of the Tīsta has been assigned at Rome to the Jesuits of Calcutta, and that to the east of the Tīsta, *i.e.*, Kālimpong or British Bhutān, to the Vicariate Apostolic of Central Bengal of the Milan Mission. The former maintain a charity school at Darjeeling and an orphanage at Kurseong, and have about 250 converts. At Pedong in the Kālimpong subdivision is the head-quarters of the Roman Catholic Mission to Western Tibet, which also has a Christian village and another church at Mariabastī, a short distance to the south. The history of this Mission is an interesting one. It was founded by Father Desgodins, who attempted in 1856 to enter Tibet through Sikkim. Repulsed by the Rājā, he determined to effect an entrance through Ladak, but on his way to Simla, the Mutiny broke out, and he was one of those besieged in the fort at Agra. Eventually he succeeded in penetrating to Western Tibet; but as it had been determined to abandon all attempts to enter Tibet from Northern India, he was recalled and directed to join the mission in Eastern Tibet near China. Here he laboured for 22 years, was imprisoned and suffered various persecutions; but finally in 1882, when it was decided at Rome to renew the attempt to enter Tibet from the south, he was allowed to establish the head-quarters of the Mission at Pedong.

According to the census statistics, the number of native Christians has increased nearly tenfold in the decade ending in 1901, rising from 298 to 2,829. The greatest success is met with among the Lepchās, who constitute nearly half the total number, and then among the Khambus and the degraded caste of Kāmis.

CHAPTER IV.

PUBLIC HEALTH.

THE district is composed of two portions, the Tarai, a low malarious belt skirting the base of the Himālayas, which is notoriously unhealthy, and the hills, where the climate is wonderfully bracing, though it must be remembered that even here the low valleys dividing the ridges are sometimes almost as unhealthy as the marshy strip at the foot of the hills. The Tarai is a tract which nature has marked out as the home of fever. Intersected by the numerous streams which debouch from the mountains, and having an exceptionally heavy rainfall, it is a water-logged region abounding in stagnant pools, which are ideal breeding grounds for the anopheles mosquito. The physical features of this unhealthy tract have been well described by Sir Joseph Hooker. "The pent-up water of the streams," he writes, "percolating their gravel beds, and partly carried off by evaporation through the stratum of ever-increasing vegetable mould, must be one main agent in the production of the malarious vapours of this pestilential region. Add to this, the detention of the same amongst the jungly herbage, the amount of vapour in the humid atmosphere above, checking the upward passage of that from the soil, the sheltered nature of the locality at the immediate base of lofty mountains; and there appear to be here all necessary elements, which, combined, will produce stagnation and deterioration in an atmosphere loaded with vapour." A tract of reeking moisture and rank vegetation, it has always been dreaded by Europeans, who used, in the days before the railway, to hurry through it as fast as they could travel, and if possible in the early morning, in order to get beyond the fatal fever zone. That such speed was necessary was proved by the fate of Lady Canning, who caught the fever which ended in her death, while halting to sketch by the road-side on her return journey from Darjeeling. The people who live in this tract have their energies sapped by fever and are far from virile. The death-rate is appalling, the average mortality being nearly 60 per 1,000 in the ten years

GENERAL CONDITIONS.

The Tarai.

I am indebted to Major F. O'Kinealy, I.M.S., Civil Surgeon of Darjeeling, for assistance in preparing this chapter.

ending in 1900, while it exceeded 71 per 1,000 in that year; and, on the other hand, the average birth-rate in the decade was only 19·4 per annum. In 1905 the death-rate was 57·70 per mille, and the average for the previous five years was 60 per mille.

There is, however, one race which inhabits this sickly region with comparative immunity, the aboriginal Meches; and the Rājbansis are also to a certain extent free from fever, largely owing to their clearing away the rank jungle round their homesteads and to the high platforms on which they erect their huts. Many of the European planters, moreover, manage to live in the Tarai without succumbing to the fevers which attack the natives, thanks to the robustness of their constitutions, to the construction of their pile-built bungalows and their situation in clear open ground, and to their careful regard of elementary hygienic rules, such as constant physical exercise, good ventilation and suitable clothing.

The Hills. In the hills an entirely different climate is met with. The damp moist heat of the Tarai disappears at the elevation of 2,500 feet, and above that level the tropical zone of fever is past. When Darjeeling is reached, the European finds a climate in many respects like that of Europe, except for its excessive humidity, which is not suitable for pulmonary and rheumatic affections. Malaria ceases to be endemic, though fevers contracted below reappear with the cold climate; and the chief danger is of having hill diarrhœa, owing to the great difference of climatic conditions and carelessness regarding diet, clothing and exercise. The change is most observable in the case of children. The thin, pallid and peevish child is not long in Darjeeling before becoming fat, rosy and active; while the child who constantly suffered from bowel-complaints or intermittent fever in the plains below becomes a different being, regaining health, strength and the restless energy of an English child. It is, in fact, small wonder that Darjeeling has for the last half century been described, even by unenthusiastic scientists, as the paradise of children.

PRINCIPAL DISEASES. Fevers. By far the greatest mortality is caused by fevers, which are generally malarial in nature. As already indicated, they are commonest in the Tarai, where they account for over 80 per cent. of the total mortality, the death-rate from fevers being over 51 per mille in 1905. The form most frequently met with is intermittent fever with enlargement of the spleen, quotidian, double quotidian and tertian being very common; but a great variety of fevers are found, including the deadly *kāla azār* and black-water fever.

Regarding the types of fever generally met with in the district, Dr. A. D. Humphry, Civil Medical Officer, Kurseong, has kindly supplied the following information. In the hills fevers are not prevalent. Mosquitoes are found widely distributed throughout the district, and are commonest during the hot season and at the end of the rains. The genera Culicidae are met with at an elevation of over 6,000 feet, and the larvae of the Anophelina genus have been seen, at the height of 4,500 feet, in a hamlet, the inhabitants of which had no history of malaria, while the children showed no enlargement of the spleen.

Malignant fevers.

The Tarai has an evil reputation for the severity of its fevers, which may be divided into malarial fevers, including black-water fever, and *kāla azār*. Simple fevers include tertian and quartan, and those cases in which the parasites are mixed. The cycle of the tertian cases lasts for 48 hours, and of quartan cases for 72 hours. The cycle of malignant fevers, including black-water fever, lasts for about 24 hours. They are prevalent all the year round, but are most common during and at the end of the rains. The malignant type of fever is especially frequent in certain localities, such as Suknā, Garidhura, Mātighara and Naksalbāri, owing to their marshy situation and to the fact that they are cramped and crowded, and have no room for expansion. Large markets are held at these places, and they are great centres for the dissemination of fevers among the smaller villages in healthier situations. A large percentage of children show considerable enlargement of the spleen, and they are the chief harbourers of the malarial parasite, a fact which points to the very wide prevalence of the disease. It is, indeed, a common saying among the Nepalese in these parts that any child born to them will not live to reach the age of two years; and the infant mortality is very great, being over 38 per cent. in 1905 for the whole of the Tarai.

Black-water fever.

Black-water fever is included under the head of malaria, as there seems to be some connection between it and malaria. Cases have been seen amongst Europeans, but not amongst natives. This fever begins like an ordinary attack of fever, ague, etc. The patient, on passing water, is astonished to find his urine of a deep black porter colour; the skin becomes coloured a deep saffron, and in the course of a few hours changes to a deep yellow, so that the sweat will stain the sheets. The temperature is variable, but may rise to 105°. The other chief symptoms are great restlessness, sleeplessness, incessant vomiting and suppression of the urine; the patient eventually dies from the blocking of the small tubes in the kidneys causing suppression, or death may be due to simple exhaustion. The exact cause of this fever has not been satisfactorily explained, as the malarial parasite is not found in all cases. That there is some connection with malaria is apparent, for it only occurs in districts in which virulent malaria prevails. One explanation which has been given is that it is caused by the improper and injudicious use of quinine. Quinine is said to bring about an unstable condition of the hæmaglobin in the blood; and, by the injudicious use of quinine, the hæmaglobin is set free in the blood and eliminated by the kidneys, producing this condition. On the other hand, it is difficult to believe that the inhabitants in certain limited areas are more injudicious in the use of quinine than elsewhere.

Kāla azār. *Kāla azār*, or as it is known locally *kāla dukh*, is an illness with characteristics resembling in many ways those of malaria. The parasite is not to be found in the peripheral blood, but in the liver and spleen, and unlike malaria, it does not live in the red blood corpuscles, but is found in the plasma and in some of the white corpuscles. Major Rogers has been able to cultivate a trypansome from patients who have this parasite: it may be mentioned here that a trypansome is found in cases of sleeping sickness. The further life history of these parasites is at present not known, or the way in which entry to the blood is gained. Sufferers from this illness complain of an irregular fever, enlargement of the spleen, and liver, loss of flesh and strength. Towards the end of the disease emaciation is very marked. The two principal features which distinguish this fever from malaria are, firstly, that quinine has no distinctive control on the fever, as in malaria, and, secondly, that the spleen is larger than in patients suffering from malaria, and does not feel soft and elastic, but hard and stony. There is little doubt that a large number of deaths returned as due to malaria are really caused by *kāla azār*.

Diarrhœa. Diarrhœa is one of the commonest diseases among Europeans in the hills, especially among those who have just arrived from the plains, so much so that it is often referred to as the hill complaint. Various theories have been put forward to account for its frequency. One popular theory is that it is due to the presence of mica in the drinking-water. It is suggested that Europeans fresh from the plains at once take more exercise, and so become more thirsty and drink more freely. This causes a certain amount of congestion of the alimentary canal; mica lodges in the follicles, causing irritation, and diarrhœa results. This theory fails, however, to account for the fact that hill diarrhœa is equally prevalent at other hill stations where the soil is not micaceous, and that the hill people of Darjeeling who drink the same water are practically immune from it.

Another theory is that it is due to the presence of organic matter in the drinking-water which is brought down from springs on the flank of the Senchal range. These springs are situated in the midst of dense forest, and there is a mass of decaying vegetation which, it is said, affects the water. Others again deny that hill diarrhœa can be attributed to any single cause, and point out that medical observation shows that it is very frequently produced by chills, errors in diet, improper clothing and overexertion, especially on first arriving in Darjeeling. It is undoubtedly most common during the rains; and it is noticeable that natives from the plains are very subject to it, while the hill people are rarely attacked.

Other diseases

Among the latter, however, intestinal worms, producing symptoms of diarrhœa, are extraordinarily common: in 1905 no less than 3,470 such cases were treated at the Darjeeling dispensary. This is probably due to the fact that they are exceedingly dirty in their habits and careless about their food and drinking-water, the latter being often contaminated in a most disgusting manner. The same cause also occasionally produces small epidemics of cholera, but these occur only at long intervals and are not serious. Phthisis is also not uncommon among the natives, owing to their thin clothing, their constant exposure to cold, damp and heat, and to their disregard of elementary hygienic laws. It is more frequent among those natives who live in solid brick stone or corrugated-iron houses than amongst those who live in wooden or wattled huts, the former being densely crowded and ill-ventilated, while the latter let in the air at every crevice. Darjeeling with its cold damp climate bears an unfavourable reputation for the treatment of consumption or pulmonary affections; but happily these diseases are not so common among the Europeans.

Goitre and deaf-mutism are fairly common. The number of deaf-mutes is greater than in any other Bengal district except Champāran, no less than 152 males out of every 100,000 males and 124 females out of every 100,000 females being returned as deaf and dumb at the census of 1901. Among other diseases may be mentioned diphtheria and enteric, of which a few sporadic cases sometimes occur; influenza, which occasionally visits the station; and rheumatism, which is commonest in the rains when it is aggravated by the excessive humidity which prevails. Fortunately, plague is so far unknown in these hills, though a few imported cases occurred amongst Europeans in 1905, one in Darjeeling town and the others in the Victoria School at Kurseong. The disease was mild, it did not spread, and there were no deaths.

VACCINATION. In the hills the people have the great horror of small-pox, and welcome vaccination. In the Tarai the sickly inhabitants are either prejudiced against it or indifferent to the protection it affords. The difference in this respect between the attitude of the people in these two tracts will be sufficiently illustrated by the fact that in the hills the ratio of those successfully vaccinated during 1905-06 to the total population varied from 68 per mille in the Kurseong subdivision to 140 in Kālimpong, while it was as low as 22 per mille in the Tarai. In spite of the lower ratio in the last tract, no less than 79·45 per mille were successfully vaccinated in the district as a whole; and in the preceding 5 years the average annual ratio was 78·87 per mille, or more than double that for the Province as a whole. As a result of these extensive operations, small-pox is a very rare visitation, and there have been only 25 deaths from it in the last three years. No other district in Bengal can show such a record of vaccination or claim such an immunity from the scourge of small-pox.

MEDICAL INSTITUTIONS. In the town of Darjeeling there are three medical institutions—the Eden Sanitarium for Europeans, the Lowis Jubilee Sanitarium for natives, and the Victoria Memorial Dispensary for natives and Europeans. **Eden Sanitarium.** The Eden Sanitarium owes its establishment to the energy of Sir Ashley Eden, Lieutenant-Governor of Bengal, who urged the necessity of providing a convalescent home for European patients from tea gardens of the district as well as for invalids from the plains. It was opened in 1883, and a hospital and operating-room have been recently added. It is the only institution in Bengal at which Europeans requiring treatment in a cool climate, and convalescents after sickness in the plains, can obtain the accommodation, care and attention they require. In one way it is a semi-charitable institution, as besides some free cots maintained by subscriptions, four free beds are maintained for the benefit of convalescent patients from the Medical College and Presidency General Hospitals, Calcutta, while poor Europeans and Eurasians can obtain admission on payment of Rs. 3 a day, a charge which does not cover the cost of maintenance, as it includes diet, accommodation, medical attendance and medicines. It is maintained at a cost of about Rs. 40,000 per annum and the annual number of admissions varies from 450 to 550; of these 300 to 400 are sick patients. It is noticeable that the institution is year by year becoming more of a hospital in which serious cases of illness are dealt with, the number of operations alone being over 60, of confinements 10, and of enteric 4 in 1905.

Lowis Jubilee Sanitarium. The Lowis Jubilee Sanitarium was opened in 1887, and is so called because it owed its establishment largely to the zeal and energy of Mr. Lowis, the Commissioner of the Rājshāhi Division, and because it was intended to commemorate the Jubilee of Queen Victoria. Its inception was due in the first instance to the liberality of Mahārājā Gobind Lal Roy of Rangpur, who offered a donation of Rs. 90,000 for some work of public utility. This sum it was decided to utilize for the benefit of Indians who had no convalescent home in the hills to which they could resort; and the Mahārājā of Cooch Behār having given a valuable site and subscriptions being obtained all over the Province, the sanitarium was eventually established. It is divided into two departments, the Hindu department for orthodox Hindus and the general department for those who have no caste prejudices. It is a most popular institution, the number of admissions being about 650 per annum. It contains 99 beds, of which 8 are reserved for females and 13 are free, and it is maintained at a cost of Rs. 25,000 a year.

Darjeeling dispensary.

The Victoria Memorial Dispensary is the Darjeeling Municipal hospital. Besides the main building containing 45 beds, there is a cottage hospital, as well as an infectious diseases hospital, and an excellent operating-room. It is a model of what mofussil hospitals should be, it is gaining popularity every year, and it is much appreciated by the people. More than 10,000 patients were treated at it in 1905, and no less than 389 operations were performed.

Other dispensaries.

In the interior of the district there are charitable dispensaries at Kurseong, where 4,000 patients were treated in 1905; at Kālimpong, where the attendance reaches the high figure of 15,000; at Pankhābāri, where an average of 2,300 per annum are treated; and at Pedong, where 6,000 out-patients are treated yearly. With the exception of the latter, these dispensaries, which are all situated in the hills, treat both indoor and outdoor patients. In the Tarai there are dispensaries at Silīgurī, Baghdogrā, Kharibāri, Naksalbāri, Phānsidewā, and Sombāri Hāt. The dispensary at Silīgurī treats both in-patients and out-patients, and outdoor relief only is afforded at the others. The attendance varies from 4,000 per annum at Silīgurī, Baghdogrā and Phānsidewā to 2,500 at Naksalbāri and Kharibāri, and 2,000 at Sombāri Hāt. Medical relief is also afforded by an itinerant Civil Hospital Assistant, who travels among the Tarai villages and visits the different *hāts* on market days.

The medical organization of Government is admirably supplemented by the Church of Scotland Mission. At Kālimpong there is a hospital, the Charteris Hospital, aided by the State, but maintained and managed by the Mission which contains 26 beds and is in charge of a medical missionary and two lady nurses. The same Mission also maintains a dispensary at Nimbong in the Kālimpong subdivision; and the extent of the work done by it in this tract may be gauged from the fact that it affords medical relief to over 17,000 patients annually. To the west of the Tīsta the Mission maintains entirely a dispensary at Kizom, a village in the Karmi estate of the late Rājā Tenduk Palger, at which 1,200 persons are treated annually. There is also a small independent medical mission at Sukiāpokhri close to the Nepalese border, and it is reported that over 10,000 persons are treated annually by the missionaries either at the dispensary or on tour. This Mission works among the Nepalese both of the Darjeeling district and Nepāl, whose only idea of curing the body is by having the evil spirit to which illness is attributed exorcised by the Nepāli *jhānkri* or sorcerer.

CHAPTER V.

AGRICULTURE.

GENERAL CONDITIONS

THE physical configuration of the district makes the conditions of agriculture in different parts exceedingly varied. The strip of country at the foot of the hills contains level stretches of alluvial soil admirably suited for rice cultivation. In the interior is a mass of hills, the slopes of which are, in places, so stony and precipitous that nothing but scrub jungle can thrive on them or an occasional tree in the crevices of the rocks. Much of the land is unsuitable for cultivation of any kind, but, on the other hand, the soil on the gentler slopes is often of wonderful fertility. The annual rainfall also varies greatly, averaging only 60 inches in parts of the Tīsta Valley, while on the outer slopes which adjoin the Duārs to the east of the district it is over 200 inches per annum. This extraordinary variety of conditions makes it practically impossible to give any connected account of agriculture in the district as a whole.

The area may, however, be divided broadly into three tracts with widely different characteristics, the mountains west of the Tīsta. the Kālimpong mountains to the east of that river, and the Tarai. At the time of its cession, the hill territory to the west of the Tīsta was almost wholly covered with forest and was very sparsely populated. Nearly all the slopes are now under tea with the exception of the portion to the north-west of the Little Rangīt river, formerly known as Chebu Lāma's grant, where native cultivation still holds its own. The cultivation of tea is, however, so predominant in the whole tract that two-thirds of the rural population are resident on the tea plantations. Kālimpong, on the other hand, contains only four tea gardens, with a total area of 10 square miles; half of the whole country is occupied by forests, and the greater part of the remainder is reserved for native cultivation, five-sixths of the inhabitants being tenants of the Government estate. Lastly, the Tarai contains a number of tea plantations along the foot of the hills, but there are also extensive areas under ordinary cultivation, and the tea-garden population is barely one-fifth of the total.

Tea, which occupies nearly one-third of the cropped area, is the all-important crop of the district, and is especially predominant in the hills west of the Tīsta; but as its cultivation is carried on almost entirely by European planters, a detailed account of it is deferred for a subsequent chapter. The present chapter will be confined to native agriculture, and as agricultural conditions differ so widely in the hills and the Tarai, those two tracts will be treated separately.

SOILS. The soil in the Tarai is composed of alluvium, a light sandy loam being most common. In the hills the greater portion of the underlying rock consists of what is known as Sikkim gneiss. The constituents of the gneiss occur in varying proportions, and the soil varies in the same relation. That most commonly met with is a rather stiff reddish-coloured loam; but its composition varies from almost pure sand to stiff red clay. Part of the area, however, is composed of the rocks of the Dāling series consisting of slates, schists and quartzites. These comprise a narrow strip of country extending from west to east along the northern boundary of the district, of another narrow strip to the south along the lower hills, and between these two of a tract about 8 miles wide on either side of the Tīsta. Further to the south is a narrow belt occupied by the foot-hills, composed of rocks belonging to the Nahun group and consisting of soft sandstones, frequently micaceous and sometimes calcareous.

The hill cultivators themselves recognize only three kinds of soil, white soil, red soil and black soil. Of these three, black soil is the richest, white soil is considered always poor, while red soil occupies an intermediate position, requiring heavy manuring to give as good an outturn as black soil. Land abounding in large rocks is usually composed of the latter soil, and is most suitable for dry crops, such as maize and *maruā*, owing to the rich vegetable mould it contains.

EXTENSION OF CULTIVATION. It was found at the settlement concluded in 1898 that there had been no increase of ordinary cultivation in the Tarai during the previous 18 years, in spite of the improvement in communications, the rise in the price of rice and the opening out of tea gardens. In fact, there was an actual decrease in the area under country crops, while there was a slight extension in the area under tea. This stationary character of the cultivation is due almost entirely to the malarious character of the climate.

In the hills, on the other hand, there has been an extraordinary extension of cultivation, especially in the Kālimpong tract to the east of the Tīsta, owing to the influx of new settlers. Here all the land available is fast being taken up, and the country has developed at a rapid rate. In certain parts it has been recognized that cultivation has already reached its extreme limit, and that its further extension is neither profitable nor desirable. Orders have accordingly been passed that no new land is to be granted in the blocks concerned, while steps have been taken to guard against the dangers of overcultivation by establishing reserves for grazing, fodder and fuel, wherever suitable pieces of waste lands are left.

The total area under cultivation in the Tarai is 70,000 acres, or 50 per cent. of the whole tract excluding the land under forest. By far the greatest portion of the cultivated area is under rice; in fact, if we deduct the area under tea, which is cultivated solely by the holders of waste-land grants, and that under thatching-grass which can hardly be called a crop, we find that 46,000 acres out of a cropped area of 54,000 acres bear this crop. THE TARAI.

As elsewhere in Bengal, it is subdivided into *āman* or *haimantik* rice, which is reaped in the winter, and *aus* or *bhadoi* rice, which is harvested in the rains during the month of Bhādra (August-September). The *āman* or *haimantik* rice is first sown broadcast in nurseries during the month of May or early in June after the first rainfall. The fields, which are situated in the low land called *rupit*, are then constantly ploughed during the heavy rains of June and July until they have become thoroughly puddled. Ridges are constructed round them to prevent the escape of the rain-water; and when the last ploughing is over, the young seedlings which are ready in the nurseries are transplanted to the fields. They are then left to mature until they are ripe and ready for harvest. The cultivator relies mainly on the copious rainfall provided by nature to fill out the ear, but irrigation is commonly practised by means of narrow water channels, locally known as *pairis*, which are led off from neighbouring streams. Rice cultivation.

The *bhadoi* paddy is grown on the higher land which is known as *faringati*. Ploughing begins in February, and is repeated five or six times. After that the land is levelled, and the weeds and clods collected are burnt, the ashes being used for manure. The seed is then thrown broadcast, and when the young shoots are four or five inches high, the fields are carefully weeded. This process is repeated when the young plants are a foot high, and they are then left to mature with the help of the rain till they are ready for reaping in August.

Next to rice, the most important crop is tea, the cultivation of which will be described in Chapter VI. It occupies 12,000 acres or 17 per cent. of the cultivated area. About 3,000 acres are Other crops.

planted with mustard and the same area with jute. The area under this valuable crop is still insignificant, but the cultivation has increased very largely since the advent of the railway owing to the greater facilities it affords for exporting the jute. Its cultivation is still extending in consequence of the high price created by the enormous demand; land which used to grow rice is now being planted with jute; and the area under the crop has increased by nearly 30 per cent. in the last 10 years. The principal other crops are oil-seeds, pulses and sugarcane, while thatching-grass occupies 4,000 acres, or 5·6 per cent. of the cultivated area.

Irrigation. Artificial irrigation is common in the Tarai, the slope of the land and the numerous small streams affording great facilities for the utilization of the water-supply. It is estimated that 60 per cent. of the low land on which *haimantik* paddy is grown, obtains the benefit of irrigation. It is most prevalent in the lands to the north and along the Mechī, Bālasan and Mahānadī rivers, where it is easy to dam up the streams and to construct water channels taking off from the nearest river or stream.

THE HILLS, *Jhūming*. Formerly the system of cultivation was that common to nomadic tribes which is known as *jhūming*. Destitute of ploughs or plough-cattle, the rude hill tribes used to burn down the jungle and grow their scanty crops on the land thus cleared. The soil soon became exhausted by a quick succession of crops raised by the hoe; and in a year or two the whole settlement would move off to a fresh patch of jungle, which they cleared and exhausted, and then deserted for fresh woods.

Terracing. With the reservation of the forests and the advent of the more skilful Nepalese cultivators, the Lepchās have abandoned this wasteful system of cultivation and have largely given up the hoe for the plough. The latter is now in universal use, except where the mountain slopes are too steep or stony for the oxen, but in places the Lepchās may still be seen turning up the soil with a rude wooden stick. From the Nepālis they have also learnt how to construct on the mountain slopes the terraces which form such a distinctive feature of Himalayan cultivation. There is no such thing as a large level field to be found in the hills; and, to allow of the irrigation which is essential to rice, terraces are cut out with great labour from the hill-side, some of them so narrow as not to admit of the use of the plough but only of the hoe. They are laid out on easy slopes within reach of some stream or spring, the water of which is ingeniously diverted into the tiny fields by means of small water channels and bamboo pipes.

Slopes. The incline of the slopes, the aspect and the elevation are all important factors to be considered by the agriculturist of the hills.

The cultivation of steep slopes and the felling of trees in such places are a fruitful source of landslips. As the roots of the trees which have been felled, rot in the ground, there is nothing to hold the soil together ; and the result is that the peasant may one day find his field has disappeared. This is particularly the case within the first few hundred feet above a river-bed, where the land is usually very steep and is partially covered with trees and brushwood even in the most closely cultivated tracts. Here the hill-side is sometimes undermined, by being cleared for cultivation, to such an extent that landslips are found extending back to the top of the ridge. If the slope is easy and not too steep, the peasant can plough the land, but when ploughs cannot be used owing to the precipitous nature of the ground, the soil has to be loosened by *kodālis* before a crop can be sown.

Aspect. The fertility of land on a hill-side also depends largely on the aspect. An eastern or southern aspect is best, as such land gets the benefit of the morning and midday sun, while a northern aspect is cold and sunless.

Elevation. The last important factor in determining fertility is elevation. Rice will not grow above 5,000 feet, and does not give a good outturn above a height of 4,000 feet. Above the latter elevation, cardamom seldom grows, while the outturn of maize, the staple food crop, is considerably less without any compensating improvement in quality. Moreover, water is scarcer on high lands, and the crops ripen later, so that a cultivator cannot get the benefit of the high prices which are obtained in the market by an early crop. The result is that the high land (*lekh*), though much healthier than the low land (*awal*), is not nearly so much sought after, as the cultivators prefer money to health. The largest outturn of maize and rice is obtained on the *awal* land, *i.e.*, the land below 4,000 feet; wheat, barley and other cold-weather crops are grown mostly on the *lekh* land, *i.e.*, the land above 4,000 to 5,000 feet; while potatoes and cardamom thrive best in land with a shady aspect, the former in *lekh* and the latter in *awal* land.

The cultivators. Lepchās mostly cultivate at the lower levels, and are particularly fond of cardamom cultivation at the bottom of the valleys, for the low elevation is that to which they are accustomed, and the comparative seclusion of the field, surrounded, as it often is by jungle, is congenial to their habits. The Bhotiās and Nepāli Gurungs prefer the higher levels, being the descendants of pastoral races who were accustomed to graze their cattle and sheep along the higher ridges; while the more careful and successful Nepāli cultivators live usually in the fertile low lands. They are by far the most enterprising cultivators, and special measures are necessary to protect the mild indigenous Lepchās from the attempts of the former to oust them from their lands. The three races may also be distinguished by their different methods of cultivation. A Nepāli will attempt to cultivate every available portion of his holding, and will keep his plough-cattle engaged as much as he can. A Bhotiā will, as a rule, keep a portion of his holding uncultivated, either to give the land a few years' rest or to allow jungle to grow for fire-wood, etc. A Lepchā will not only leave a part of his holding out of cultivation, though not so much as the Bhotiā, but will also cultivate a portion by hand with a small spade and a spike instead of with the plough.

Irrigation. The only crops which receive artificial irrigation are cardamom and the rice grown in terraces, for which a plentiful water-supply is an absolute necessity; while maize and other crops depend, as a rule, on the rainfall. The cardamom fields, being generally near the beds of the larger streams in the valleys, are irrigated by channels leading off from them; while the irrigation channels for rice are taken from streams and springs on the hill-sides wherever they can be found.

PRINCIPAL CROPS.

Maize. Maize is the staple crop in the hills, and the cultivators mainly depend on it for their livelihood; they are in fact contemptuously called the maize-eaters by the wealthier inhabitants and townsfolk. It thrives between elevations of 1,000 and 7,000 feet above sea-level, and it will grow on almost any soil in the tracts reserved for cultivation. Black soil is usually preferred for the cultivation, but red soil, if heavily and frequently manured, gives excellent results. It grows well even in rocky places, as the rocks prevent the black soil with its rich vegetable matter from being washed down the hill-side. The seed is sown in the latter half of February or March, when good showers of rain are required to enable the young plants to spring up. Sunshine, with intervals of moderately heavy rain, is then necessary to bring them to maturity, and in the ordinary course of affairs the crop is ready for harvesting in August or September. It is a crop which is often damaged by bears and elephants, especially in the neighbourhood of the forests; and the other danger to which it is exposed is that of being carried away by land-slips on the steeper slopes.

Rice. Rice is a valuable crop in the hills, and is almost invariably a wet crop grown on irrigated land (*pānīkhet*), while maize is raised on unirrigated land (*sukhākhet*). On steep slopes the labour of making the narrow terraces is very great, but as the site of a rice-field is always selected so that it can be irrigated from some stream, the crop is a certain one and amply repays the labour expended. The seed is sown in May in a small nursery which can be easily watered, and by the time the rains break, it is ready for transplantation to the little terraced field. Heavy rainfall is then required in order to fill the streams from which the rice is irrigated, and if there is a sufficiency of water, it comes into flower in October and is ready for cutting in December or January.

Cardamom is the most valuable crop grown in the hills. It is raised on the rich black mould at the bottom of the valleys or on the sides of some hill stream, and will not grow above the height of 5,000 feet. A good water-supply and moderate warmth are all required for its proper development, besides a sufficiency of shade, which is secured by planting trees at suitable intervals. The irrigation of the crop entails a great deal of labour; the fields are hot and wet; and the work is consequently unhealthy. Moreover, lying as they do almost in the beds of the streams, the fields are very liable to land-slips, and in every year of heavy rain many are washed away, or the irrigation channel is destroyed, thus reducing the value of the land to that of an ordinary maize-field. The plant grows in the form of a bush with large dark leaves. There is practically no production for the first three years; but after that time the outturn increases till the plant is seven or eight years old. It then diminishes, but will continue in decreasing quantities till the fifteenth year. The plant flowers in May and is reaped in August and September, the fruit being cut from the bottom of the stem and then dried in a kiln. **Cardamom.**

The principal cold-weather crops are wheat, barley, mustard and buck-wheat, which are frequently sown in land which has already yielded a crop of maize. Wheat and barley are grown only by the Lepchās and Bhotiās, who eat the pounded flour with their tea. Mustard is grown chiefly by the Nepālis for the sake of the oil which is indispensable to their households. The buck-wheat, the scarlet flower of which lends a vivid touch of colour to the bare hill-side, is grown by all the hill races. **Other crops.**

Among other crops may be mentioned potatoes, tobacco, *kodo* (*Paspalum scrobiculatum*), various millets and pulses, and sugar-cane, which is grown in small plots near the cultivators' homesteads. The last crop of any importance calling for separate mention is *maruā* (*Eleusine Coracana*), from the seed of which the mild native beer called *chhang* is made, as well as from the *kodo* millet seed. It is sown in May and reaped in December, and grows well on land situated at a high elevation. Generally the land on which *maruā* has been grown does not produce any other crop in the same year.

FRUITS AND VEGETABLES. A large variety of fruits is grown in the valleys and on the slopes where the rainfall is not excessive, such as plums, peaches apples, pears, mangoes, plantains and oranges. The peaches are hard and somewhat bitter, not fit for eating but only for stewing, the plums are almost equally bitter, the pears and apples are also rather hard and flavourless, but the oranges are of a delicious flavour and large quantities are exported. Pine-apples, both English and Singapore, are common, as well as the jack-fruit, guava, *papaya* and several kinds of lemons. Another common fruit is the tree tomato which was imported from America. It resembles an egg in shape, and has all the characteristics of a tomato, except for a slight but pleasant acidity. It seems possible that large fruit plantations might be successfully started on some of the slopes where there is plenty of sunshine and the rainfall is not heavy, but so far European enterprise has not been attracted to the scheme.

The vegetables are those of a tropical as well as of a temperate zone, the commonest being the potato, of which large quantities are raised by the natives. It is somewhat surprising that a large potato-growing industry has not sprung up in these hills, where the conditions are in many respects so favourable; but hitherto blights have proved very destructive and have checked the efforts made to extend its cultivation. Among other vegetables the commonest are brinjals, sweet potatoes, chillies, garlic, onions, pumpkins and yams. A large number of English vegetables have been introduced and are grown successfully, such as rhubarb, tomatoes, turnips, cabbages, cauliflowers, beans, peas, beet-root, carrots, parsnip, leeks and celery, while mint, parsley and thyme are common garden herbs.

CATTLE. In the Tarai the common domestic animals do not differ from those found elsewhere in Bengal, oxen being used for agriculture, buffaloes reared for their milk and for sale, and goats and pigs reared for food. In the hills there are two special breeds of mountain cattle, the Siri and the Nepāli, a cross between the two being called Siri-Kutcha. The former are large and rough-coated, and the latter small and smooth-skinned; both breeds are good climbers and thrive in the forests. The Siri and Siri-Kutcha are especially powerful animals, which are largely used for cart traffic. The Sharpā Bhotiās from Nepāl, as well as the Bhotanese Bhotiās, graze large herds of cows and buffaloes; while the Nepalese cultivators use a few bullocks for their ploughs or for transport. When these are no longer fit for ploughing or cart work, they are generally sold or slaughtered for food. The Bhotiās, being a pastoral race, are the chief graziers, and keep a large number of cows, even when they take to cultivation. The buffaloes are not used for agricultural purposes; the females are kept for milk, while the males are slaughtered for food. Pigs are common, being most numerous in the parts which are largely inhabited by Lepchās and Limbus. The Nepalese Gurungs graze large flocks of sheep, taking them to the heights in the rains, and in the cold weather bringing them down to the plains for sale.

The small but sturdy breed of Bhotiā ponies, introduced from Tibet and Bhutān, is well known. They are remarkably coarse-bred animals with upright shoulders, ugly heads and great bone; they are valued for their sure-footedness and great endurance, and are used both as pack animals and for riding. Mules are imported from Tibet to Kālimpong, where Government buys them for transport. Small but sturdy donkeys are also bred in the Kālimpong tract. **Ponies.**

The principal pasture grounds are the reserved Government forests. In the cold and hot weather the lower ranges are used, but, as the rains approach, the scourge of leeches drives the cattle and sheep up to the higher mountains between 10,000 and 12,000 feet. For the herds of the professional grazier, which are too large for village grazing, the reserved forests, as a rule, afford the only pasturage available. The plough cattle of the cultivators, and such milch cows as they keep for their daily use, find pasturage in the cultivator's own fields, the maize fields being utilized for two or three months after the maize is cut, and the rice stubble after the rice is reaped. But after the maize is in the land again, and until it is reaped, *i. e.*, from March to September, the need of grazing as near as possible to his homestead is strongly felt by the cultivator; and to supply this want village grazing grounds have now been reserved in the Kālimpong Government estate. **Pasturage.**

Veterinary aid is afforded by an itinerant Veterinary Assistant, paid by the Darjeeling Improvement Fund, his duty being to tour in the interior and deal with outbreaks of epidemic disease. Altogether 2,400 cattle were treated in 1903-04, 1,500 in 1904-05 and 1,300 in 1905-06. The diseases most common are rinderpest and foot-and-mouth disease; there were 800 cases of the former and 1,400 of the latter in 1903-04. **Veterinary aid.**

As regards the special breeds of cattle in the hills, Lieutenant-Colonel F. Raymond, F. R. C. V. S., Superintendent, Civil Veterinary Department, Bengal, has kindly furnished the following account:— **Breeds of cattle.**

Darjeeling has two ordinary zebu breeds of cattle, which go by the names of Siri and Nepāli or Pahāria, and a cross of these two is called the Siri-Kutcha. There are a few tame specimens of the Methun or Gayal to be found, and some crosses between it and the Siri breed, but they are very few in number.

Siri breed. The word Siri comes from the Sanskrit word *Srī*, and signifies that the breed is handsome and has good qualities. The usual colour is black, black and white, or red and white. The bulls have well-developed humps with a tuft of hair growing on it. The cows have only rudimentary humps. In height, the bulls measure up to 54″ behind the hump and the cows up to 50″. They are remarkably well suited to the high lands of the Himālayas, and withstand the cold and wet, leeches and other hill pests in an extraordinary manner, and are able to feed themselves by grazing, if turned loose in the forests. As workers they do well, a pair being able to drag about 8 maunds of material in light carts up to about 12 miles a day on fair roads. They are also very useful as milkers, a well-fed cow giving about 6 seers of really excellent milk daily.

Nepāli breed. The Nepāli or Pahāria cattle are considerably smaller than the Siri breed and are good climbers. The bulls stand only about 45″; the cows are rarely over 40″, generally less. They only give about 2 seers of milk. Though they are hardy animals, they are essentially poor men's cattle; and whenever the owner finds an opportunity to do so, he crosses his Nepāli cow with a Siri bull.

Siri-Kutcha breed. The Siri-Kutcha is the result of a cross between a Nepāli or Pahāria cow and a Siri bull. The cross is always a great improvement upon the Nepāli, and most of the cattle which are seen working in the district are Siri-Kutchas of sorts. They are however not in any way as good as the pure bred Siris.

Methun breed The Methun (*Bos frontalis*), also known as Gayal, are really wild cattle in the Bhutān, Burmese and other jungles; a few have been caught and tamed by Bhotiā and Indo-Chinese tribes, and some few, in a domesticated condition, can be found in Darjeeling. They are generally a deep chestnut or brown in colour, but they look black in the forests. Their legs are white; the dewlap is broad and heavy, and the anterior dorsal shins rise considerably, causing the dorsal ridge to be prominent. The bulls are about 58″ high and the cows about 50″; they are heavily built, short-legged and very powerful. They become very docile when tamed. The cows give 3 to 4 seers of very rich milk. From time to time, the Gayal has been crossed with the Siri, and it is said the result has been satisfactory. Though the quantity of milk has been somewhat diminished, the quality has considerably improved; and the power and endurance of the progeny has notably developed. The people, however, generally object to crossing the Gayal with the Siri for the same reason that Hindus refuse to breed mules.

Attempts have been made from time to time to import English and Australian cattle into the hills; but they promptly die if allowed to graze; or they fall down the *khuds*. Hence they have to be stall-fed and treated as exotics. The result, as far as is known, has been discouraging and unprofitable.

CHAPTER VI.

THE TEA INDUSTRY.

PROGRESS OF THE INDUSTRY. THE establishment of the tea industry in Darjeeling is due to the enterprise of Dr. Campbell, who was appointed Superintendent of Darjeeling at a time when attention was being attracted to the possibility of starting and developing the cultivation and manufacture of tea in the territories under the East India Company. In 1834 the Governor-General, Lord William Bentinck, had appointed a committee "for the purpose of submitting a plan for the introduction of tea culture into India." This committee was apparently ignorant of the fact that in 1821 Major Bruce, and in 1824 Mr. Scott, had discovered the tea plant growing wild in Assam; and much expense and considerable delay were consequently incurred in bringing plants and seed from China, and importing Chinamen to teach the people of India how to grow the plant and manufacture tea. Satisfied that a great future might lie before the industry, Government itself undertook the formation of experimental plantations in Upper Assam and the districts of Kumāon and Garhwāl; in 1839 private speculation took the field, and the Assam Tea Company was formed.

Introduction of tea. In 1840 Dr. Campbell was transferred from Kātmāndu to Darjeeling, and there started the experimental growth of tea. It was soon found that the plant throve readily at this altitude, and others began to follow Dr. Campbell's example, seed being distributed by Government to those who desired to cultivate the plant. Writing in 1852, Mr. Jackson says in his Report on Darjeeling*—"I have seen several plantations in various stages of advancement, both of the Assam and China plant, and I have found the plants healthy and vigorous, showing that the soil is well adapted for the cultivation. In the garden of the Superintendent, Dr. Campbell, in Darjeeling, in the more extensive plantations of Dr. Withecombe, the Civil Surgeon, and Major Crommelin, of the Engineers, in a lower valley called Lebong, the same satisfactory result has been obtained: the leaves, the

* Selections from the Records of the Bengal Government, No. XVII.

blossom and the seeds are full and healthy; the reddish clay of the sides of the hill at Lebong seems to suit the plant better than the black loam of Darjeeling. This has been the result at and about Darjeeling itself, at a height of 7,000 feet; but the opinion of Dr. Hooker and of others competent to judge seems to be that there is too much moisture and too little sun at Darjeeling to admit of the cultivation on a large scale becoming remunerative: this objection, however, does not apply to the lower sites of Pankhabari and Kurseong, where a plantation of both tea and coffee has been established by Mr. Martin, and the plants are now in a highly-thriving condition. In this tract of country, between the Morung and Darjeeling, every variety of elevation and aspect is to be found, and there seems to be little or no doubt that tea cultivation in that tract would answer."

These plantations appear to have been merely experimental plots, but by the year 1856 the industry began to be developed on an extensive scale, especially on the lower slopes, as it was believed that the elevation of Darjeeling was too high for the plant to be very productive. According to the account of a contemporary writer,* tea plants had been sown and raised by the end of that year at Takvār to the north by Captain Masson, at Kurseong by Mr. Smith, at the Canning and Hope Town plantations by the Companies attached to those locations, by Mr. Martin on the Kurseong flats, and by Captain Samler, the Agent of the Darjeeling Tea Concern, between Kurseong and Pankhābāri. At the same time, Government endeavoured to supplement the efforts of these pioneers of the industry by distributing several maunds of tea-seed among the native cultivators. The writer doubted, however, whether the erratic disposition of the Lepchā would allow him to wait four years before he reaped the fruits of his labour; and the event has proved the correctness of his judgement.

Early tea gardens.

The year 1856† may accordingly be taken as the date at which the industry was established as a commercial enterprise. In that year the Alubāri tea garden was opened by the Kurseong and Darjeeling Tea Company, and another on the Lebong spur by the Darjeeling Land Mortgage Bank; in 1859 the Dhutaria garden was started by Dr. Brougham; and between 1860 and 1864

* Darjeeling, by the Revd. T. Boaz, LL.D., Calcutta Review, Jan. 1857.

† In the Annals of Indian Administration (1862) it is said: In Darjeeling the first trial of the tea plant was made in 1841 with a few seeds grown in Kumaon from China stock. It was not till 1856 that the first plantation was started at Kurseong, and another at Darjeeling, by Captain Samler, who was also the first to grow coffee.

four gardens, at Ging, Ambutia, Takdāh and Phubsering were established by the Darjeeling Tea Company, and the gardens at Takvār and Badamtam by the Lebong Tea Company. Other gardens which were started at this early period were those now known as the Makaibāri, Pandam and Steinthal tea estates. All these estates are situated in the hills, but about this time the planters began to turn their attention to the Tarai, where experimental plantations had already been started. Here, in 1862, the first garden was opened out at Champtā, near Khaprail, by Mr. James White, who had previously planted out the Singel estate near Kurseong, which is still one of the largest gardens in the district. Others followed suit, and by the end of 1866 more gardens had been opened out in the Tarai.

Extension of cultivation.

In the meantime, however, the development of the industry in the hills had been even more rapid as the suitability of the soil and climate to the growth of tea became apparent; Government offered land to investors on favourable terms; and the industry rapidly developed. By the end of 1866, *i.e.*, only ten years after the establishment of the industry on a commercial basis, there were no less than 39 gardens with 10,000 acres under cultivation, and an outturn of over 433,000 lbs. of tea. In 1870 there were 56 gardens with 11,000 acres under cultivation, employing 8,000 labourers and yielding nearly 1,700,000 lbs.; and in 1874 the number of gardens had increased to 113, the area under cultivation to 18,888 acres, the outturn to 3,928,000 lbs., and the labour force to 19,000 souls. In other words, between 1866 and 1874 the number of gardens under tea was almost exactly trebled, the area under cultivation increased by 82 per cent., while the outturn of tea was multiplied nearly ten times. Since that time the industry has progressed steadily until no less than 50,600 acres, or 79 square miles are under tea cultivation. The following table illustrates the advance which has been made during the last 30 years:—

YEAR.	Number of gardens.	Area under cultivation in acres.	Outturn of tea in lbs.
1874	113	18,888	3,927,911
1885	175	38,499	9,090,298
1895	186	48,692	11,714,551
1905	148	50,618	12,447,471

From this table it will be seen that during the last ten years there has been but little extension of cultivation owing to the fact that all the land available and suitable for tea within the area reserved for it has now been taken up, while the number of gardens has been reduced in consequence of the amalgamation of several estates. Of the 148 gardens now in existence, 71, with an area of 25,800 acres under tea, are situated in the Darjeeling thāna, which includes the Kālimpong hills to the east of the Tīsta. The latter portion of the district is however almost entirely closed to tea, both because the greater part of the tract is a forest reserve, and because most of the remainder has been reserved for native cultivation. Nearly 30 square miles have been reserved for tea, but the land, as a rule, is so barren and precipitous, and so unsuitable for the growth of the tea plant, that, notwithstanding the eagerness for grants of tea lands, little of it has been taken up, and the few gardens being worked, which are on the lower slopes adjoining the Duārs, comprise only 10 square miles. Of the remaining tea gardens, 46 with an area of 16,900 acres under tea are situated in the Kurseong thāna, *i.e.*, on the lower hill slopes; and 32 estates with an area of 7,900 acres under tea lie within the Silīgurī thāna, *i.e.*, within the Tarai. The development of the industry in the latter portion of the district has had to contend against serious drawbacks. It is an extremely unhealthy tract, it has suffered severely from blights, and it has been further handicapped by having its labour drawn away to the Duārs. The result is a tendency for the Tarai tea estates to fall into the hands of native managers and owners, while elsewhere the industry is almost entirely under European supervision and management, and is supported by European capital.

Present position of the industry.

In addition to the 50,600 acres actually under tea, which in themselves constitute one-third of the total area under cultivation, the tea estates in the district include 49,300 acres which have been taken up by planters but have not been planted with tea. The total area occupied by the land leased out to the various tea estates is thus approximately 100,000 acres, or a little less than one-seventh of the total area of the whole district. The industry is now the staple industry of Darjeeling, and its importance may be realized from the fact that no less than one-third of the entire population reside on the tea gardens and that the manufacture and cultivation employ a labour force amounting, according to the census of 1901, to 64,000 coolies. At the same time, there is not much more room for further extension, as in the area reserved for tea cultivation almost all the land capable of being planted with tea has already been taken up. The only extension possible appears to be that gradual slow extension which goes on from year to year within the areas of the different gardens wherever labour and land are available for clearing an additional few acres. It is true that the district contains large areas on which tea could easily be grown, but they are either covered by valuable forests or are in the possession of native cultivators; and the land occupied by the latter has been so impoverished and bared of forest that it would require a long rest before it would be suitable for tea.

Its prospects. Since the year 1897 the industry has undergone a period of severe depression; the prime cause being overproduction, which was brought about by reckless extensions in India, Ceylon and Java, following upon a time of high prices and great prosperity. Other factors have united to render the position more acute; notably the depreciation in the value of the rupee since the standard of exchange became fixed, and the crushing increase in the duty on tea which has been imposed in Great Britain, though considerable relief has lately been given in regard to the latter. The natural expansion of the tea trade, the opening of new markets on the Continent, and the steady supplanting of the Chinese article in the Australian, American and Russian markets has now resulted in the demand having overtaken the rate of production, so that the prospects of the industry are distinctly favourable. The most promising feature in the situation is that in future the crop from Ceylon is likely to show a decrease, rather than otherwise, because much of the area under tea in that island is being rapidly interplanted with rubber trees.

CULTIVATION.* Most of the area in Darjeeling has been planted with the China variety, which was for many years considered the only kind suited for the production of fine tea. Some planters of experience still hold to this view, though it is now a very rare thing for the pure China plant to be planted. Of late years the variety known as the "Assam indigenous" has been much in favour, and it is certainly capable of producing the very finest tea; but it is very delicate, and with anything like rough treatment soon becomes so weak as to be unremunerative. A good hybrid from these two varieties above mentioned has proved most suitable all round. Some fields have been planted with the "Manipur indigenous" which is the most hardy of all the varieties, and gives a good yield, but the tea produced is almost invariably coarse and rank in flavour. These three are the principal varieties of tea at present cultivated; but recently a distinct variety

---

* I am indebted to Mr. Claud Bald of Takvār for the account of the Manufacture and Cultivation of Tea. For fuller details, Mr. Bald's monograph "Indian Tea" (Calcutta, 1908) may be consulted.

has been found in Formosa, of which some specimen plants are now being raised.

The most important factors in the production of tea of good quality are soil, weather, blights and management. Being a mountainous district, Darjeeling contains many varieties of soil, but that which has proved most suitable for the growth of tea is a loamy soil well supplied with nitrogen. Those estates which produce the finest quality have, almost invariably, at least some fields which contain a chocolate-coloured ferruginous soil, which, on analysis, is found to contain a high proportion of phosphoric acid and potash. General conditions.

The rainfall varies greatly in different parts, owing to the configuration of the district, ranging from 150 inches a year on the slopes facing the plains to 70 inches in some of the valleys about 20 miles north. There is a similar difference between the temperature at different elevations, tea-fields at the bottom of the valleys having a tropical temperature like that of the plains; while it is only 2 degrees above that of London at the height of 7,000 feet, which is within the zone of the plant. Other things being equal, the tea at lower elevations produces the larger crops. There is a great profusion of leaf in the hot damp heat of the lower valleys and the Tarai, but, on the other hand, though the crop is smaller at higher elevations, the quality is distinctly better.

Darjeeling tea is famous for its peculiarly fine flavour; but the quality produced varies greatly in different parts of the district, and varies also remarkably at different seasons of the year on the same estates. It is indeed not uncommon for teas produced in one month to sell for double, and occasionally four times, the price of teas produced on the same estate two months later or earlier, as the case may be. The finest teas are usually produced from the second growth, just before the advent of the monsoon rains, and again at the end of the season, when the growth has become slow and the sap thick. It is a generally recognized fact that the teas produced during the rainy season are watery and poor.

As regards the pests and blights which attack the tea plant, their name is legion; and it will suffice to mention only the red spider, the mosquito blight and the green fly. The red spider first appeared in 1876 in the dry, hot valley of the Little Rangīt, but is now general in the Tarai and the hills. Nowhere has it been eradicated, but an effectual remedy is found in the application of sulphur. The mosquito blight has long done serious damage in the Tarai and the Mahānadī valley, and of late years has carried its ravages into the hills. The green fly is looked Blights.

upon with some favour, because although it reduces the crop to some extent, its action is such as to produce conditions favourable to the development of good quality in the tea.

Seed.

It is a well-known fact that when plants of one variety are raised in the immediate neighbourhood of an inferior class, the seed of both becomes more or less hybridized, owing to insects carrying pollen from one flower to another during the flowering season. For this reason, it has been customary to establish special fields for seed trees, in remote places by themselves, where the finer class of plants are supposed to be out of reach of hybridizing influences. The seed trees in such cases are not used for producing leaf, but are allowed to grow naturally, without pruning, so as to yield the largest crop of seed possible. Some planters, however, have objections to this plan, chiefly because the Assam indigenous grows to a height of 25 to 30 feet, and when attacked by blights is very difficult to treat; while the blights are almost certain to be communicated to their progeny by means of the seed. When the seed trees are kept to a moderate height by pruning and plucking, the crop of seed is very small, but it is proportionately good. The importance of selection, in the case of both seed and transplants, cannot be overrated. The seed from plants which have proved to be not only high-class, but also blight-resisting, is sure to give the best results; and this is the only seed fit for propagation.

Preparation of land.

Any good deep soil is suitable for tea, and the presence of some stones does not do harm, provided there is a preponderance of good soil. It has been found that poor land does not pay for its cultivation, while a shallow soil, overlying rocks, is fatal to success. The reason for this is that the tea plant develops a long tap root, which requires to penetrate to a depth of at least 4 or 5 feet, from which depth it extracts moisture during seasons of drought. Flat land does not always prove best; and the finest soil is usually on land with a moderate slope, the surface containing a good deal of humus, the substratum a chocolate-coloured sandy loam, and the sub-soil a sandy clay. The slope may be anything up to an angle of 40 degrees; but an angle of even 45 degrees is not uncommon, although the steeper lands are more liable to have the soil washed away during heavy rain, and are also exposed to the serious risk of great blocks being completely carried away by land-slips.

The first step in the preparation of the land for tea is to clear away the trees and shrubs. This is best done by grubbing them out from the roots: when half uprooted, the weight of the tree itself completes the work. Trees, stumps, and bamboo roots are ther rolled down the hill-sides into the ravines, where they can be chopped up and carried away to the factory for fuel at leisure. It is always good policy to give a thoroughly good digging to the land at the start. Some planters dig to a depth of 18 inches all over, grubbing out stumps and roots of every kind; and at this stage, many planters form the land into terraces, while others prefer to form the terraces gradually after the planting has been done. This is one of the most important works in the garden, as its prosperity and even its very existence depends on the stability of the soil, which otherwise is washed away by the torrents of rain to which it is exposed

The land having been prepared in this way, germinating beds are formed for seed, and nurseries are made at different points, wherever water is available, on sites not too far from the land to be planted. The soil in these nurseries has to be thoroughly pulverized, and all jungle, roots and stones removed. Beds, 4 feet wide, are then thrown up on fairly level land with shallow drains, one foot wide, between. Sometimes the seed is sown directly into the nursery beds; but if the quality is at all doubtful, it has to be germinated in a hot-bed. Good seed is usually sown about one inch deep and at a distance of 4 inches apart. As soon as the plants have sprung up, and their individual character indicated, the unsatisfactory ones are uprooted, including all those of an inferior class or habit, and all weakly or blighted seedlings, only the fittest being allowed to occupy the ground. Nurseries.

After the land has been dug all over and the soil thoroughly pulverized, it is necessary to put stakes in the ground at the points where the plants are to be. This is a task calling for no little ingenuity and care, as it is necessary to arrange the plants in straight lines up and down the steep hill-sides and across the rounded slopes and ridges, and at the same time to plant them out in the terraces without loss of space or overcrowding. The distance between the plants on hill-gardens averages 4 feet by 4 feet (base measurement), a plan which gives 2,722 plants to the acre. Some estates, however, are planted much closer, and in some instances as close as 3 feet by 1 foot, but this crowding is a distinct and serious drawback after the estate comes into full bearing. The pure Assam indigenous gives the best return if planted out at distances of not less than 5 feet by 5 feet. Much of the early planting in Darjeeling was done on the system called "seed at stake," the seeds being taken from the germinating beds as soon as they sprout, and placed in the ground, one or two at each stake. Planting out seedlings is done during the early Planting.

months of the monsoon, and is a proceeding which requires very great care in order to ensure complete success. The most successful work is done during the cold season, when the plants are taken up very carefully, each with the soil round its root adhering in a hard unbroken ball.

Culture. Many of the older estates have deteriorated to an alarming extent, owing to destructive methods of culture, especially persistent digging during the currency of the monsoon rains. The effect is particularly noticeable on some ridges and knolls, from which the soil has been completely washed away to a depth of a foot or more; the roots of the bushes standing out of the ground to that extent. Some "wash" is, of course, unavoidable under any method of culture, where the soil has to stand the onslaught of tropical rains, and where the land is at all steep. The reflection that each inch of rain means a fall of 101 tons of water to each acre, ought however to give pause to the energetic cultivator, who is stirring up the hill-side and exposing the soluble and best constituents of his soil to the full force of the flood. Fortunately, better methods are now being followed, digging being done only during the dry season, and hand-weeding largely resorted to during the rains; while the faces of the terraces and the spaces between the lines are kept free of weeds by sickling.

Manuring. Manuring in Darjeeling has never got beyond the experimental stage. Unlike the Chinese, the native cultivators of India will not handle natural manure. All manure from stables and byres is made full use of, but cannot go far. In some instances, the old thatch from the cooly-lines is carefully gathered up for use on the tea-fields, its value as a manure proving quite remarkable. Artificial manures have been tried, but are too expensive, the cost of carriage by the hill railway being prohibitive. Top dressing with new soil from the forest is by far the most effective and permanent method of improving the land, and those estates are fortunate which have such soil available in the ravines or on patches of uncultivated land near enough for the purpose. A covering of 6 inches deep requires 650 tons of soil to the acre. Mulching with leaves from the forest is not expensive, and is of great value. All varieties of leguminous plants and trees possess more or less the capability of fertilizing the soils in which they grow, owing to the action of certain bacteria which are fostered on their roots. This fact is now being realized, and on many estates trees, such as *Albizzia*, *Erythrina*, *Dalbergia*, etc., are being planted within the tea. Leguminous crops are also grown on some estates for green manuring, the most favoured being certain varieties of Phaseolus (*dāl*).

Pruning. When allowed to grow naturally without pruning, the China plant grows to a height of about 15 feet. It branches right from the ground, forming a shrub of rather elegant appearance. The Assam variety is more like a tree in form, with a clean stem springing for some feet up from the ground, and grows to a height of 25 and even 50 feet. For the purpose of the planter, it must be kept in the form of a low bush, not higher than 3 feet, although 2 feet is the most useful height. The bush has also to be trained so as to have a good spread of plucking surface. Pruning begins one year after planting, when the young plant is cut to a height of 6 or 8 inches, and sometimes lower, as a bushy growth from the very root is desirable. A year later the pruning is carried to a height of about 12 or 14 inches; and thereafter a little higher year by year, all moribund wood being cut out, and unnecessary twigs removed. About once in fifteen or twenty years, the bushes require heavy pruning in order to remove the snags and knots formed by successive light prunings. The life of the tea bush, under favourable circumstances, extends to about 100 years.

Plucking. After the spring has set in, the bushes are allowed to develop on each shoot a growth of five leaves, besides the terminal bud. The pluckers then come along, and with forefinger and thumb nip off a piece from the top of each shoot, consisting of the bud and two leaves, leaving three leaves on the bush. The latter are left because the bush requires new leaves by which to breathe and maintain its health. Only young tender leaves are fit for making into good tea, and the old-matured and coarse leaves are useless for the purpose. The bud makes what is called "Orange Pekoe" and "Broken Orange Pekoe"; the young leaf next it is "Pekoe"; and the coarser leaf is "Pekoe Souchong."

After the first plucking, new buds start away from the nodes or eyes at the base of the remaining leaves. As these develop to a growth of three leaves, the new bud and two leaves are again plucked, leaving only one leaf; and so on throughout the season. A growth of shoots is termed a "flush." The shoots are plucked only as they become ready, and the whole estate is gone over once a week, the interval becoming longer as the season advances, till it is once a fortnight at its close. Plucking begins at the end of March, and closes about the end of November. The average crop is about 240 lbs. of tea per acre; but the crop from some estates is as low as 160 lbs., while on one or two estates it has reached fully 480 lbs. per acre. Each bush is plucked about thirty times in the course of the year, and produces a fraction under two ounces of tea per annum. When it is considered that it requires 4 pounds of green leaf to produce

1 pound of manufactured tea, and that the total outturn of the district is over 12 million pounds, it will be realized what an enormous number of bushes have to be plucked and what a large labour force must be employed. At the same time, the pluckers are so quick and skilful at their work that it is often impossible to follow the motion of their hands as they pluck the leaves.

MANUFACTURE.

Withering.

The leaf having been plucked is taken to the factory, where it is weighed, and the process of manufacture then begins. The first process in the factory is withering, which means that the leaf is simply left thinly spread upon trays or frames for ten to twenty hours. By the end of this time it becomes soft and flaccid, having lost about one-third of its weight by the evaporation of moisture. During this process the enzyme within the sap of the leaf becomes considerably developed and increased, if the withering is carefully regulated. In some cases this is done artificially by dry heated air which is drawn over the leaf by means of large revolving fans.

Rolling.

The next process is rolling, the prime object of which is to bruise the leaves, so that the cells become ruptured, and the sap brought to the surface. The leaves are thus converted into a wet mass, and incidentally the leaves become curled and twisted. The use of machinery in this process has marked a very important advance upon the objectionable method of rolling by hand, still followed by the Chinese, which involves the addition of a good deal of perspiration to the leaf during manipulation.

Fermentation.

Fermentation begins as soon as the sap becomes liberated in the process of rolling, and is due to the sap coming into contact with the atmosphere. At one time it was thought that the change which takes place could be described as simply "oxidation"; but the careful analytical researches of Dr. Mann have established the fact that the prime agent is an enzyme within the sap itself, which becomes active as soon as it comes into contact with the atmosphere. It is properly described as fermentation, but is quite different from the fermentation which produces alcohol; it is not a process of decomposition or putrefaction and it has no connection with bacterial action. During fermentation, the leaf is spread about two inches deep upon low tables, or on beds of cement or glass. In this process it assumes the coppery colour which infused tea leaves are known to have. It takes about five hours to complete, reckoning from the time that the leaf enters the rolling machine. The temperature during the process must not be over 80 degrees (F.).

It is important that the drying or firing should begin exactly when the fermentation has reached the proper stage. Drying is almost invariably done by machinery, of which there are many kinds in use. By means of these machines, the leaf is passed through a chamber filled with air which has been previously heated by passing through the tubes of a huge stove. The temperature of the air is kept at about 240 degrees; and the leaf is slowly turned over within the machine while drying. The operation requires great care, as all the moisture must be eliminated, and yet the tea must have no trace of scorching, which destroys its delicate flavour. **Drying.**

After firing, the tea is sifted by machinery into the various grades already enumerated, according to the size of the leaves; and the manufacture is complete. **Sifting.**

It then only remains to pack the tea in the well-known chests of commerce; and this is done on most of the large estates by means of machinery. The tea is invariably packed in close-fitting chests, lined with lead, which has to be carefully soldered up, as tea is very susceptible to moisture, and must be hermetically sealed in order to retain its flavour and aroma. **Packing.**

A large number of the tea factories have been built in situations where full advantage can be taken of the water-power available in the mountain streams on the estates. There is always considerable difficulty in keeping such waterworks running smoothly during the rainy season, when the streams are apt to swell into uncontrollable torrents, occasionally overstepping their usual bounds, and causing damage even to the factory buildings. In one instance, a factory which had been erected too near the stream was completely carried away by an accident of this sort. During the last decade or so, the old water-wheels have all given place to the more modern and effective turbine as a driving motor. On many estates water power is not available at any point where a factory could safely be erected. In such circumstances, the transmission of power by means of electricity from turbine to factory is certain to be adopted in course of time. Already one or two estates have gone in for the system, and others are deterred only by a want of sufficient technical knowledge of electric machinery. A rapid transformation within a few years is probable, because the district has already become seriously denuded of timber forest, to an extent which some of the older residents regard as somewhat alarming, and as calculated to damage in many places the stability of the hill-sides. Each estate of 1,000 acres cultivation consumes not less than 15,000 maunds, or 500 tons, of wood fuel each year. Some factories are now supplied **Water-power and electricity** with coal from the plains; but the cost of carriage renders it expensive. Where tea machinery is run by water-power, the bill for fuel has been reduced by about two-thirds.

LABOUR. The coolies employed on the estates in the hills are almost entirely Nepalese, and in the Tarai they are drawn both from Nepāl and Chotā Nāgpur. The Nepālis, who form the great majority, although extremely improvident, are a cheerful, hardworking, and enterprising race, courageous to a degree, and pleasant to work with, so long as they are treated with fairness and consideration. The average rate of wages for men is Rs. 6, for women Rs. 4-8, and for children Rs. 3 and Rs. 2; but in addition to wages, they get quarters on the garden, often with water laid on and free medical attendance and medicine. The necessaries of life are few and cheap; so that, granted a little thrift, the cooly can save a considerable portion of his wages. Of late years, there has been considerable stringency in the labour market, owing to many interests competing with the planters for the labour available. This is a matter of vital importance to the industry, as cheap labour is essential to its prosperity. The labour is free, *i.e.*, it is not controlled by any special legislative enactment, as in Assam There are no agreements, labourers being free to come and go as they like. They are employed directly by *sardārs*, who take advances from the estates on account of the coolies' debts, which are many and heavy: some estates occasionally suffer heavy losses on this account owing to the managers having no proper hold upon the debtors. A cooly earning Rs. 6 a month usually spends about Rs. 50 upon his wedding, and occasionally a similar amount upon a funeral, all borrowed money. The cooly looks to the *sardār* for an advance, and the *sardār* to the manager. The latter lends without interest; and if he will not, the cooly and *sardār* turn to the Mārwāri, the Shylock of the hills, who commonly charges 75 per cent interest. In that case their financial slavery is assured, and they rarely escape the toils of the money-lender.

MANAGEMENT. Except in the Tarai, almost all the managers and assistants on the estates are Europeans. It is a remarkable fact that, though educated natives are much cheaper than Europeans, it has not been found economical to employ them generally, although here and there a few natives have done remarkably well, and have proved themselves worthy of full trust in positions of responsibility The result is that although the industry in the hills is now fifty years old, it is still almost entirely in the hands of Europeans.

As regards the buildings on a tea estate, the factory, the manager's bungalow and the cooly-lines, the following account* may be quoted: "In the past the planter has been his own architect and clerk of works, planning, estimating for, and vigilantly superintending every operation from the sawing of the timber in the forest, the making and burning of the bricks to the completion and full equipment of the edifice. The buildings have generally to be designed to suit the sites after costly excavation on the steep hill-sides, and in the case of amalgamation of factories, for instance, ingenuity has to be exercised in the erection of a main central building to connect two others at different elevations, the ground floor of the one being near, but not on an identical level with the first floor of the other. The whole has to be designed for convenience, to deal with increased crops, and meet modern requirements. The managers' bungalows are generally charmingly situated, little being required, in the shape of landscape gardening, to make the surroundings altogether beautiful. These buildings, with their white-washed walls and shining roofs studded over the hill-sides, surrounded by the green tea gardens, present to the traveller at all points a striking and pleasant feature of the landscape. Many of the dwellings of the garden-labourers are substantially built with brick walls, corrugated-iron roofs at the higher, and thatch for coolness at the lower elevations, and verandahs in front. The floors are well raised, and space, convenience, comfort, and sanitation are provided up to and, in most instances, in advance of the tastes and wishes of the occupants. Few of them would tolerate fire-places for the exit of the smoke by chimneys in the walls; but on account of warmth, or for other reasons best known to themselves, prefer it finding its way from the centre of the floor through the ventilators in the roof. Many darken or block up entirely the windows provided for them. However, the modern cooly dwellings are a very great improvement upon the huts once universal, and in most cases all is being done for the coolies' comfort that is practicable or at present wise." THE BUILDINGS

In conclusion, a word may be said about the planter, the man on whom the responsibility of management rests. His duties are multifarious, including the supervision of the cultivation, the control of the manufacture, the management of the large labour force employed, the construction of roads in the estate, and often the erection of the buildings. He must therefore combine, as far THE PLANTERS.

* Tea Planting in Darjeeling, by G. W. Christison, Journal of the Society of Arts, June 1896.

as possible, the knowledge and skill of an agriculturist, engineer and architect, and even, to some extent, of a doctor; and above all, he must have firm control over his labourers, the art of management, and generally the power of conduct. The planters are, as a rule, considerate masters, anxious to promote the welfare of their employés; and in public life they form a community prompt to devote their time and energies to the public service and to the development of Darjeeling. In the hills they constitute the agency for the construction and repair of roads, the establishment of schools and the improvement of sanitation; and, in the words of a former Lieutenant-Governor of Bengal, they are the backbone of the British government in the district.

CHAPTER VII.

FORESTS.

THE most remarkable feature of the forests of Darjeeling is the wonderful variety of species that they contain; there are, in fact, probably few places in the world in which so many different types of forest exist within so small an area. At the highest altitudes, from the elevation of 8,000 to 12,000 feet, forests of silver fir (Abies Webbiana) are found interspersed with grassy slopes which are dotted here and there with the whitened stems of dead trees. These gradually merge into extensive rhododendron forests, which are at present useless from an utilitarian point of view, owing to their remoteness and inaccessibility, but present scenery of a very picturesque character. Lower down are thickets of bamboos, which at 6,000 feet give way to forests of chestnut, maple, oak, magnolia and laurel, the chief source of supply of timber and firewood to the station of Darjeeling. Still lower down the oak disappears, and the chief species are maple, birch, alder and the graceful *pipli* (Bucklandia populnea). At the level of 4,000 feet these trees of the temperate zone are replaced by mixed forests, which, with the exception of the *tun* (Cedrela toona), are of little value except for fuel. At 3,000 feet the upper limit of the *sāl* (Shorea robusta) is reached, and this tree is the chief constituent of valuable forests in the lower hills, extending from the extreme west of the district to the Chel river on the east, where it ceases abruptly, probably owing to a change in the geological formation. Besides the *sāl* forest, the Tarai, as the lower foot hills and adjacent plains are called, contains swamp, river-bed and savannah forests. These grassy savannahs used to be a great source of danger to the adjoining forests from their extreme inflammability, but, owing to the effectual system of fire protection introduced by the Forest Department, they are gradually disappearing with the advent of tree growth. GENERAL DESCRIPTION.

With the exception of 20 square miles of forest in the Kālimpong tract under the charge of the Deputy Commissioner, all the forests in the district are reserved forests controlled by the Forest Department. They cover an area of 445 square miles, ADMINISTRATIVE DIVISIONS.

and the greater part are situated at elevations above 6,000 feet and below 3,000 feet, these being the limits adopted when forest conservancy was first started in the hills, as being the actual upper and lower boundaries of native cultivation, and generally of tea cultivation also. They are divided for administrative purposes into three divisions, the Darjeeling, Kurseong and Tīsta divisions, the head-quarters of which are at Darjeeling, Kurseong and Kālimpong respectively.

The forests of the Darjeeling division occupy the summits and slopes of four hill ranges, and, generally speaking, are situated above 6,000 and below 3,000 feet, the ridges and slopes in the intermediate areas having been taken up for tea cultivation or ordinary native crops. The four ranges under forest are the Singalīlā range stretching southwards from Kinchinjunga along the western boundary of the district; the Ghum range which runs due west from Ghum till it meets the latter range, of which it may be regarded as a continuation; the Senchal-Mahaldiram range, which trends generally southwards towards Kurseong; and the Takdāh range, which branches off from the main Senchal ridge and extends in a north-easterly direction to the junction of the Tīsta and Rangīt rivers.

The forests of the Kurseong division are partly in the hills and partly in the Tarai. In the northern part of the division, they are scattered over the hills up to an elevation of 6,500 feet, the greater portion lying below 3,000 feet. The most extensive areas of forest, however, lie along the foot of the hills between the Tīsta on the east and the Champtā river on the west. The other forests in this division lie to the west and south-west of the latter, and form a series of blocks scattered over the neighbouring hills and the Tarai as far west as the Mechī river, which forms the boundary between the Darjeeling district and Nepāl.

The Tīsta division comprises the reserved forests situated on the mountain slopes to the east of the Tīsta river, where they form a continuous belt extending round that portion of the district. East of the Chel river, they extend northwards up to the frontiers of Sikkim and Bhutān; and west of that river, they extend in a narrow strip along the southern boundary of the district westwards as far as the Tista, and then northwards up the left bank of that river to where the Rangpo constitutes the frontier of Sikkim. They are also connected with the broad tract of forest east of the Chel river by a narrow belt of forest running from Paiengaon to Laba, where they enlarge northwards up to the Sikkim frontier, thus forming a complete circle. Besides this, a narrow stretch extends from Laba some 12 miles in a south-westerly direction into the interior of the circle thus formed.

A short account* is given below of each division, and the following table shows the average annual outturn, revenue and expenditure during the last ten years :—

Division.	Timber.	Fuel.	Minor produce.	Revenue.	Expenditure.
	Cubic feet.	Cubic feet.	Rs.	Rs.	Rs.
Darjeeling ...	177,330	986,650	8,100	91,560	72,860
Kurseong ...	203,770	371,480	2,820	55,900	38,300
Tista † ...	128,630	630,900	7,800	34,980	36,600
Total ...	509,730	1,989,030	18,720	1,82,440	1,47,760

The reserved forests of the Darjeeling division extend over 117 square miles and fall naturally into three classes:—(1) the semi-tropical low level forests up to the height of 4,000 feet; (2) the temperate forests between 4,000 and 9,000 feet in elevation; and (3) the sub-alpine forests which are found above the height of 9,000 and up to the elevation of 12,000 feet. The zone between 3,000 and 6,000 feet is principally taken up by tea and other cultivation, and the irregular patches of forest which encroach on it from above and below are therefore reckoned with their more important neighbours. DARJEELING DIVISION.

The semi-tropical low level forests are found in the valleys and clothe the lower slopes of the hills. *Sāl* (Shorea robusta), which is found in patches up to about 3,000 feet in elevation, is the most valuable as well as the most characteristic timber tree in this zone, and special measures are taken to develop it. The other valuable species in order of importance are the *auli chāmp* (Michelia Champaca), *lampāte* (Duabanga sonneratioides), *pāka-sāj* (Terminalia tomentosa), *pānisāj* (Terminalia myriocarpa), the rubber tree (Ficus elastica), corrupted by the Nepālis into *lābar*, and the *tamā-bāns* (Dendrocalamus Hamiltonii). This zone is included in the Tīsta Valley range, and extends over an area of 12 square miles. It stretches from the Rayeng river along the western bank of the Tīsta as far as the junction with Semi-tropical forests.

* The account of the Darjeeling and Kurseong Divisions has been contributed by Mr. H. King Robinson, Deputy Conservator of Forests, and that of the Tīsta Division by Mr. Tinne, Deputy Conservator of Forests. Mr. W. H. Lovegrove, Officiating Conservator of Forests, Bengal, has kindly revised them. Fuller details can be obtained from the Working Plans.

† The revenue from the sale of sleepers in the Tīsta Division is included in the accounts of the Kurseong Division.

the Great Rangīt, and then follows the western bank of the latter river as far as the junction with the Little Rangīt. The ground is nearly everywhere steep and sometimes precipitous, and is broken up by several enormous landslips. The valley has an unenviable reputation for malaria, and is considered even more unhealthy than the Tarai. Consequently, nearly all work has to be suspended during the rains, as the labourers are hillmen, who are particularly subject to the ravages of fever at the lower levels.

Temperate forests.

The temperate forests are characterized by excessive atmospheric moisture, and their most striking feature is the wealth of orchids, ferns, mosses, and other epiphytes and creepers, which load the branches of the older trees; while the large number of tree ferns, interspersed with small bamboos and wild plantains, which are found in the lower part of this zone, add to the beauty of the scenery. The trees, many of which grow to a great height and girth, belong to a great number of different genera, the principal being oaks, magnolias, chestnuts, and maples, though laurels are also well represented. The most valuable are the *safed chāmp* (Michelia excelsa), the timber of which is used for panelling and flooring in houses, and is consequently in great demand; the *tun* (Cedrela toona), which furnishes one of the best light planking woods in India: it was formerly the most popular wood for tea boxes, but it is now getting expensive, and is more largely used in house building: the *būk* (Quercus lamellosa), which is used for heavy beams, and supplies the best fuel; the *phalāt* (Quercus lineata), which yields a wood like *būk* but slightly inferior in quality; the *singari katūs* (Quercus pachyphylla); the *katūs* (Castanopsis hystrix); the *piplī* (Bucklandia populnea), and the *okar* or walnut (Juglans regia), a fast-growing and hardy tree, which yields a wood equal in quality to English walnut. Extensive plantations of all these species have been and are being made to assist the natural re-stocking of those areas over which fellings are being made, and about 50,000 young plants are annually transferred from the nurseries to the forest.

Sub-alpine forests.

The sub-alpine tract is characterized in its upper portion by a forest of silver fir (Abies Webbiana) which lower down gives place to the hemlock spruce (Tsuga Brunoniana). This forest is mixed with birch (Betula utilis) and an undergrowth of various rhododendrons, and in the upper parts the yew and juniper are noticeable. The undergrowth in these forests between 7,000 to 10,000 feet consists, as a rule, of an almost impenetrable thicket of hill bamboos (Arundinaria racemosa), which are used for making mats. At about the height of 11,000 feet several species of aconite are found, which are of some economic importance. Sheep and goats crossing over from Nepāl are always muzzled to prevent them eating it, as animals strange to the place are killed by the poison, if this precaution is not taken. The sheep that live in the neighbourhood have however become accustomed to the plant, and eat the young green leaves with impunity, though they are careful to avoid the old leaves which are richer in poison.

Forest development. The forests in this division were the first in Bengal to come under the management of the Forest Department. The first attempt at forest conservancy was inaugurated in 1864 in the forests near Darjeeling, which then extended over 15½ square miles and were situated for the most part on the main Senchal range between Ghum and Kurseong. Since that time the area of reserved forests has gradually expanded, the largest addition being made in 1882-83 when the Singalīlā forest, extending over 66 square miles, was purchased from the descendants of Chebu Lāma, to whom this tract had been granted. They now form an exceedingly valuable asset, not only because of the vast amount of excellent timber they contain, but because they are the main source from which the station of Darjeeling obtains its supply of fodder, fuel and milk. Excluding the forests in the Tīsta Valley and on the Singalīlā range, three-fifths of the total area have been thrown open to grazing, and it is perhaps not too much to say that if it were not for the facilities of pasturage thus afforded, Darjeeling would be forced to import its milk from the plains.

KURSEONG DIVISION. For purposes of general description, the Kurseong division may be divided into three parts, the Tarai, the lower hill forests and the middle hill forests. They are situated between the height of 200 and 6,500 feet on the outer slopes of the Himālayas in the southern part of the district, and occupy a total area of nearly 107 square miles.

Tarai forests. The Darjeeling Tarai, the southern part of the division, is a slightly undulating or level tract of country from 200 to 1,000 feet in elevation, intersected by numerous streams and having the Tīsta river for its eastern and the Mechī river for its western boundary. The area of the reserved forests in the Tarai is about 45 square miles, consisting of a compact block to the east and of smaller detached areas to the west. The only well-represented tree of great commercial importance is *sāl* (Shorea robusta), which extends over an area of 29 square miles. Occasionally almost pure, it is generally mixed with a large proportion of other species, of which the most noticeable are *tatre* (Dillenia pentagyna), *kambi* (Careya arborea), *baherā* (Terminalia belerica), *goelo*

(Callicarpa arborea) and *udāl* (Sterculia villosa). The areas in the Tarai which do not contain *sāl* chiefly consist of old river-beds, in which the principal species are *simal* (Bombax malabaricum), *khair* (Acacia Catechu), *sisu* (Dalbergia Sissoo) and *safed siris* (Albizzia procera).

Lower hill forests. The lower hill forests, stretching from the Tarai up to a height of 3,000 feet, contain about 51 square miles of reserves. Here again the most important tree is the *sāl*, but it is generally restricted to the ridges, the intervening depressions being usually occupied by scrub and creeper jungle, bamboos, usually Dendrocalamus Hamiltonii, and a great variety of trees of other species, such as *chilauni* (Schima Wallichii), *lampāte* (Duabanga sonneratioides), *goguldhup* (Canarium sikkimense), *simal* (Bombax malabaricum), *chāmp* (Michelia champaca), *tun* (Cedrela toona), *mallata* (Macaranga denticulata), *pānisāj* (Terminalia myriocarpa) and *phallida* (Erythrina sp.). The rubber tree (Ficus elastica), though rare, is indigenous along the foot of the hills.

The Bāmanpokhri block, consisting partly of Tarai and partly of lower hill forest, deserves special mention. By nature intended for *sāl* forest, 477 acres were planted with teak, and about Rs. 40,000 were spent over it. The largest trees are now 4 feet in girth, but no seedlings have appeared. If left to itself, it would soon be mostly suppressed by *sāl*, which thrives in this tract. As an experiment in planting exotics, it is so far a success, but these costly experiments at the expense of species of equal value are now recognized as unnecessary, and do not form part of the forest policy in vogue.

Middle hill forests. In the forests above 3,000 feet many of the species noted above as occurring in the depressions in the lower hill forests are the principal trees, but towards the upper limit birch, *vern. sauer* (Betula cylindrostachys) and alder, *vern. utis* (Alnus nepalensis) are noticeable. Above 6,000 feet various species of Quercus, Echinocarpus and Machilus are characteristic. The crop varies from tangled scrub and creeper jungle, with bamboos in places, to fully-stocked high forest.

Forest development. The forests of the Division were selected out of unoccupied waste at the disposal of Government and gazetted as reserves, mostly in 1879 and 1880, but the Paglājhorā block, which had been included in a grant to a private person, had to be bought back in 1885 to protect the hill cart road and railway from landslips. Since the forests have been properly protected, a very great improvement has been effected. In the Tarai, where fires used to be frequent and the high forest was being replaced by savannah with *sāl* scattered in it, there is now hardly any grass to be found, and the *sāl* is most promising and very dense. The only disappointing block is Dhobijhorā, nearest to Kurseong, which has been overfelled and overgrazed in the past.

When the Forest Department took charge of these forests, they had to allow advanced fellings in the principal *sāl* areas in order to meet the demand of the railway for sleepers. When this ceased, the fellings remained unregulated, except to a certain extent in the Bāmanpokhri block and the Suknā forests, until the introduction of Mr. Hatt's working plan in 1904. These unregulated fellings were confined to the more accessible areas, but as the possibilities were seldom worked up to, no damage was done, and the forest, though still immature, had time to recover. Mr. Hatt's working plan includes all the areas in the Division, and the systematic management therein provided will not only allow of a proper interest being earned from the forest capital, but will also provide for the proper treatment of the young trees now growing to maturity.

Forest management. The railways are the markets for sleepers, and the tea gardens for box planking and for a limited supply of firewood. In order to prove to the contractors that *sāl* could be profitably put on the market, Government had recently for a few years to do its own exploitation. This is now generally being replaced by sales of standing trees to purchasers who are working them out by their own agency. Box planking for the tea gardens is converted by the agency of the managers. A certain amount of timber and fuel from the neighbouring hill forests is sold in Kurseong for local consumption, and a little charcoal is also made there. In the lower hill forests rubber (Ficus elastica) is being planted out to a moderate extent below Latpanchor and in Bāmanpokhri, in places where *sāl* does not grow. This should in time yield a handsome revenue from forests which now give absolutely no return.

Forest protection. Fire protection has been attempted since 1875-76 by clearing a wide belt along the boundaries, and by fire-tracing the principal roads on each side. This has been so successful that very little inflammable grass land remains inside the forest, and as a rule the only kind of fires to be feared are leaf fires, which may occur in the hot weather, and, if helped by high southerly winds, may extend over considerable areas. In existing circumstances they would not do serious damage, but precautions are taken to prevent them. Cattle-grazing is light in the hills and practically non-existent in the Tarai, and the damage done in this way is slight, except in the Dhobijhorā block, where grazing has to be allowed for the milk-supply of the Government schools.

Wild elephants do a certain amount of damage in the forest, but this is not to be compared to the destruction they would cause if there were young plantations in the Tarai. The chief objection to them is that they endanger the lives of every one working or living in or near the forest. Tigers are plentiful, but as man-eaters are practically unknown, work is not interfered with by them. Man is a far greater source of danger, from his tendency to light fires and from his inability to understand that he has not the same claim to all forest produce as he has to air and water. The result is that thefts are matters of almost daily occurrence, wherever there are cultivators living near the forest. In the Tarai, where the inhabitants are mostly Santāls and Rājbansis, the principal petty offence is shooting game. In the hills, which are peopled by Nepālis and Bhotiās, the damage done to forest growth by the illicit cutting of poles, trees and fodder is very considerable. This is worst in the Dhobijhorā block, where the surrounding country is most thickly populated, and where protection with a view to future developments is most important.

TISTA DIVISION. The Tīsta division includes the reserved forests in the Kālimpong subdivision to the east of the Tīsta, the total area being 221 square miles, or 52 per cent. of the total area of the subdivision. These forests form belts along the left bank of the Tīsta river and the northern boundary of the Jalpaigurī district, where they descend to 500 feet elevation; and they occupy the greater portion of the higher ground of the subdivision, *i.e.*, those parts which range from 5,000 feet to 10,300 feet in elevation. In the greater part of the tract the rainfall is very heavy, amounting to over 200 inches a year on the southern foot hills, and the soil and sub-soil, consisting for the most part of soft micaceous schist, are easily eroded.

Lower forests. The most important forest tree at the lower elevations, *i.e.*, up to about 3,000 feet, is *sāl* (Shorea robusta), but it is not found east of the Chel river, and west of that line is generally confined to the ridges, from which it shows little tendency to spread. The number of other kinds of trees which grow at the lower elevations is very great, but the most important from an economical point of view are *tun* (Cedrela toona), *lampāte* (Duabanga sonneratioides), *chāmp* (Michelia champaca) and *mallagiri* (Cinnamomum cecidodaphne). Two small patches of *naugesuri* or iron wood (Mesua ferrea) are found in the plains portion of the division, and the oldest trees were probably planted round and mark the site of former Buddhist temples.

Upper forests. At what may be called the intermediate elevations, *i.e.*, between 3,000 feet and 5,000 feet, few trees of much importance occur. *Tun* is the most valuable, but *chilauni* (Schima Wallichii) is most abundant. Above 5,000 feet the trees characteristic of the Darjeeling forests such as *pipli* (Bucklandia populnea), *okar* or walnut (Juglans regia), *safed chāmp* (Michelia excelsa), *būk* (Quercus lamellosa), *katūs* (Castanopsis hystrix), etc., predominate, but at the highest points forest trees disappear, and the ground is almost completely covered with a dense growth of maling bamboo (Arundinaria racemosa). *Singari katūs* (Quercus pachyphylla), three or four species of rhododendrons, and the Himālayan hemlock (Tsuga Brunoniana) are characteristic of the over-wood up to 9,500 feet. Many parts of the forests, especially at the lowest and middle elevations, appear to have suffered severely from *jhūming* and fires in the past, and in such places the ground is often covered by a dense growth of creepers and worthless shrubs.

Forest development. These forests were reserved during the years 1879-82, when the remainder of the subdivision was reserved for native cultivation. Most of the ridges and the tops of the hills were reserved to protect them against erosion, and other parts were retained in order that they might serve as permanent timber and fuel reserves and as grazing grounds for the tenants of the estate, or because the ground was too steep and rocky to permit of cultivation. For many years after reservation, as the tree-covered waste outside the reserves was more than sufficient to supply local requirements, and as the forests were practically inaccessible to outsiders, very little timber or firewood could be sold from them; but, later, the construction of the Tīsta cart-road supplied a means of extracting the most valuable kind of timber, *i.e.*, *sāl*, at least from the most accessible parts of the forests on the left bank of the Tīsta; whilst the development of the tea industry in the Jalpaigurī district, accompanied as it has been by the destruction of outside forests, has caused a demand to spring up in the eastern part of the subdivision which borders on that district. But in the absence of roads or of supplies of timber of valuable kinds, the high level and intermediate forests are still practically unworkable.

The bulk of the forests are, at present, only protected to prevent erosion on a large scale, such as would certainly occur, and would equally certainly result in the devastation of large tracts of cultivable land in the plains below, if the forests were destroyed, as the land for the most part is subject to heavy grazing, and, outside the forests, to shifting cultivation. The existing supply of trees is also preserved against the time when the development of the surrounding country and the improvement of the roads will make the profitable extraction of timber possible. Though the Government ryots of the subdivision require comparatively little timber and firewood, and have up to the present been able to supply their wants from the wooded portions of the subdivision which have not been included in the reserves, sooner or later a considerable part of them will have to depend on the reserves for such supplies, and the station of Kālimpong is in the same way becoming a local market of some importance. Up to the present the inhabitants of the subdivision have used the forests chiefly for grazing and for obtaining supplies of cut fodder. The bulk of the demand for fuel comes from the tea estates in the Western Duārs, and though in some cases managers have found it more economical to use coal, still the position of their gardens makes others more or less dependent on the fuel obtainable locally; and as the supply on their own grants diminishes or gives out altogether, they are obliged to turn to the reserves to supply the deficiency.

Forest management. Grazing is regulated by the rules laid down for the Darjeeling district in 1884, by which about half the area of the forests is open to grazing, not more than one head of cattle being allowed for every ten acres of the open area. Some very large areas in the upper forests have not, up to the present, been grazed over,* but in certain cases the effects of overgrazing are very noticeable. In times of scarcity, a large part of the population, especially the Lepchās are able to support themselves entirely on the produce of various plants (roots, fruits, etc.) growing wild in the forests, and the removal of all such jungle products is allowed free of charge on the understanding that they are not for sale.

Almost the whole of the yearly yield of *sāl* timber is extracted in the form of metre-gauge railway sleepers, for which the Eastern Bengal State Railway provides a practically unlimited demand, the remainder being cut into sleepers for the Darjeeling-Himālayan Railway, or into scantlings, which are carted down to Silīgurī and disposed of there.

Protection. Owing to the heavy rainfall and the long rainy season, and to the evergreen nature of most of the forests, the danger from fire is not great. The *sāl* forests, which are most liable to serious injury, are fortunately placed in this respect, as a large proportion of them are situated in most inaccessible positions. The fire season lasts from the 15th February till the 31st May; but no serious fires have occurred within the last ten years. In the forests bordering on the Jalpaigurī district wild elephants are a

* No graziers will drive their cattle up to the Rishi-lā owing to the fact that a species of pieris grows there, which is very deadly to cattle.

very serious nuisance. They trample down and destroy young trees which have been planted out at considerable expense, and by terrorizing the coolies engaged in work in the forests, make the successful management of some of the most important parts of the forest a matter of considerable difficulty.

DEFORESTATION. There is a belief generally current in the district that deforestation has been carried out to such an extent that the distribution of the rainfall has become less regular than it used to be, and that the danger caused to the stability of the hill-sides by such denudation is so great as to imperil the prosperity of the district. As regards the latter point, there can be no doubt that the wholesale clearance of forest is extremely dangerous in a land of steep valleys like Darjeeling. In many parts the rainfall is very great, the slopes are steep, and the soil and sub-soil are generally friable and easily eroded. In fact, in large areas the slopes cannot maintain themselves unless they are protected by trees, shrubs and undergrowth. Where they are bared of forest and scrub, the rain rushes down with irresistible force, carrying the soil away, suddenly swelling the rivulets and torrents, undermining the hill-sides and causing disastrous landslips. Conditions are such that even a forest covering does not give absolute protection from erosion, especially when the forest is grazed over to any great extent. On steep slopes, if the foot of the hill once slips away, the slope becomes still steeper, and the hill does not lie at a natural angle of repose. Nature will then continue the wash-down from above until the natural angle is obtained. The fact, however, remains that, when such slopes are under forest, the natural angle of repose is steeper than if there were nothing but bare land. Moreover, in a well-kept forest the trees and undergrowth act somewhat like a sponge, allowing the rain to reach the ground and percolate gradually, and erosion is so slow that no great harm is caused. It is a very different matter when the forest covering of steep slopes is destroyed, and the ground is subsequently subjected to rough cultivation or heavy grazing. In such cases, the heavy rain ploughs through the exposed soft soil, landslips soon begin, and, unless their occurrence stops cultivation and grazing, continue until the slope finds a level at which it can rest. This process may last a century or more. Even where the land has a gentle slope, though no great harm is likely to be caused to the land itself by the clearance of forest, the rain-water flowing off it quicker than it would if it was under forest, is likely to do great damage by causing floods lower down.

There is no question that large stretches of forest have been cleared away since the British first came into possession of the country. All the early travellers are unanimous in describing the district as entirely covered by forest. Captain Herbert, then Deputy Surveyor-General, spoke of the mountains in 1830 as "completely clothed with forest from the very top to the bottom," and General Lloyd in 1837 similarly described it as "clothed from the top of the hills to the very bottom of the valleys with a dense forest." This however was only a natural incidence to a country which was practically uninhabited. With the phenomenal increase of the population which has taken place, and with the establishment of the tea industry, it was necessary to clear the land in order to support the people and to allow of the cultivation of the tea plant. The result has been that, within certain limits, the forests have yielded to the plough and settled cultivation, and that elsewhere they have been ruthlessly swept away by the planter. In the words of Colonel Waddell,* "whole forests have been annihilated, leaving here and there only a solitary tree or narrow belts of trees in the ravines, as evidence of the magnificent woods which have fallen a sacrifice to advancing cultivation."

When the hill territory was first acquired, the officers of Government and the settlers were impressed by the great extent of the forests rather than by the benefits to be derived from them. For many years their sole aim was to expedite their conversion into cultivated fields, and reckless exploitation ran riot. This wholesale destruction of the forests was put a stop to by the Forest Department, which inaugurated forest conservancy in this district in 1864, the forests round Darjeeling being the first to be reserved in the whole Province. It was high time, as the forests within five or six miles of the station had already been almost entirely cleared of trees fit for timber, reckless waste prevailed, and valuable timber trees were indiscriminately cut for firewood along with the inferior species. The altered appearance of the hills round Darjeeling sufficiently attests the valuable work of regeneration and conservation which has been carried on by the Forest Department since that time. Similar work has been done in other parts of the district; and the forests now under the management of the Department cover no less than 445 square miles or 38 per cent. of the total area of the district—a fact which may fairly be regarded as a proof that denudation cannot have taken place to any very serious extent. In the hills, these forests extend over the ridges above 6,000 feet and the valleys below 3,000 feet, while the slopes between those levels are covered by tea estates occupying 100,000 acres.

---

* Among the Himalayas.

Nowhere have the forests been so effectively cleared as on the gardens, but there is no proof that the rainfall has been in any way affected thereby. The records of the Meteorological Department do not afford grounds for the belief that the distribution of the rainfall has been materially altered during the last 20 years, and the figures quoted in the margin for the station of Darjeeling do not disclose any diminution of rainfall as compared with half a century ago.

Year	Rain in inches.	Year.	Rain in inches.
1851 ...	126·50	1901	109·73
1852 ...	104·70	1902	148·55
1853 ...	118·18	1903	92·71
1854 ...	147·33	1904	121·53
1855 ...	108·22	1905	153·97
Average ...	120·98	Average	125·30

The other great source of danger lies in the Khās Mahāls or Government estates, especially those lying to the east of the Tīsta. Here the forests have been widely cleared owing to the influx of settlers and the advance of cultivation. When this tract was annexed from Bhutān in 1865, it was estimated to have a population of 3,536 souls, but the volume of immigration has been extraordinarily large, and the number of inhabitants has steadily risen from 12,683 in 1881 to 26,631 in 1891, and finally to 41,511 in 1901. The population has thus more than trebled within the last 20 years, and every year there has been a steady reclamation of waste land, which has not always been properly controlled. The bulk of the new settlers have made their homes in lands where population was sparse or which were scattered in tracts of forests. Consisting for the most part of the poorer and more remote blocks, these tracts have been the last to attract settlers, but they are now eagerly taken up by those who can find no good land elsewhere, to the consequent destruction of the former forest. More than half, however, of the 400 square miles which constitute the Government estate is occupied by reserved forests; only 85 square miles have been brought under cultivation, including 10 square miles under tea; and the forests that are left still cover 20 square miles. Government has, moreover, directed that no new cultivation is to be allowed in a large number of blocks, and rules have been laid down to regulate the felling of trees. But in some parts the mischief has already been done, notably in the basins of the Chel and Lish rivers, which drain into the Duārs. The clearing and grazing of these areas must, if not checked, eventually result in serious damage to the underlying tracts, owing to the erosion which sets in; but these tracts are for the most part situated beyond the borders of the district.

In conclusion, reference may be made to the danger caused by the deforestation of Sikkim and Bhutān * in consequence of the practice of *jhūming* or shifting cultivation. This practice destroys the scrub jungle and herbaceous growth over large areas, which are then roughly hoed or ploughed for cultivation. Heavy rain falling on the loosened earth on the steep hill-sides, washes the soil away, rendering the area more or less barren, and in the majority of cases causes landslips, which may, and often do, cause great damage to life and property. The evil does not end here, for perhaps its most serious aspect is the destruction caused in the plains by such rivers as the Tīsta, Jaldhākā, Torsā, Raidāk, Chel, Sankos and others. The heavy rain of the hills falling on the *jhūmed* surfaces of the steep slopes runs off very much more rapidly than nature ever intended it to do, with the result that an enormous head of water reaches the main streams suddenly, instead of gradually, and causes abnormal floods. Such floods have been not infrequent in recent years, and there is reason to suppose that their unusual violence is due to the denudation of the country they drain. This however is not the last or greatest injury caused by such deforestation; for the brunt of the damage falls on the plains below, into which the rivers debouch, and not on the countries in which the evil of *jhūming* is most prevalent. During these floods the water of the rivers is heavily charged with the soil and clay removed by the rain from the hills above. On reaching the plains, where the current becomes slower, this silt is deposited, with the result that the river-beds become raised, until a point is reached when the river has to change its course. Any one familiar with the Duārs is aware of the thousands of acres of sandy unprofitable old river-bed due directly to this cause; and a further example of the damage caused in this way is afforded by a stream running into the Tīsta about 10 miles above the Tīsta bridge, which has brought down silt to such an extent on the Sikkim side that it has caused that river to alter its course to the west bank, where it is now engaged in rapidly cutting away the most valuable *sāl* forest of the whole Darjeeling district.

* This account of the results of deforestation in Sikkim and Bhutān has been prepared from a note by Mr. J. W. A. Grieve, Deputy Conservator of Forests.

CHAPTER VIII.

NATURAL CALAMITIES.

THE greatest danger to which the district is exposed is its liability to landslips. This is no new feature in the Lower Himālayas, for more than half a century ago Sir Joseph Hooker mentioned several cases of enormous landslips during his travels. In one case, the whole face of a mountain appeared more or less torn up for a mile in extent, and was constantly shifting from the action of streams and from the continued downward sliding of the hill-side. In another place a landslip, of which he witnessed the traces, had carried away acres of rock soil and forest; while another immense landslip, which extended fully 4,000 feet up, seamed a whole mountain side, moving every year, and shooting down masses of mud and rocks. "The most prominent effect of the steepness of the valleys," he wrote, "is the prevalence of landslips, which sometimes descend for 3,000 feet, carrying devastation along their course: they are much increased in violence and effect by the heavy timber trees, which, swaying forwards, loosen the earth at their roots and give impetus to the mass. This phenomenon is as frequent and destructive as in Switzerland, where, however, more lives are lost, from the country being more populous, and from the people recklessly building in places particularly exposed to such accidents. The fragments of rock precipitated are sometimes of enormous size, but being a soft mica-schist, are soon removed by weathering. It is in the rainy season that landslips are most frequent, and shortly after rain they are pretty sure to be heard far or near."

No great calamity is, however, known to have been caused in this way, so far as the district of Darjeeling is concerned, prior to the year 1899, when the unprecedented rainfall which occurred during the 23rd, 24th and 25th September, following an already excessive seasonal rainfall, resulted in disastrous landslips, which caused the loss of many lives and the widespread destruction of houses, roads and property. The immediate cause of the disaster was a cyclone which burst upon the district on the 23rd September. After entering the district, the storm took a

LANDSLIPS OF 1899.

north-westerly direction and was most severe from a point 6 miles south of the station of Darjeeling, extending to the north-east as far as the Rammān and Rangīt rivers and to the north-west as far as the boundaries of the district. In its severest form, it did not extend eastwards of the Tīsta river, and it is noticeable that a rainfall of only 13·96 inches was recorded at Kālimpong, and of 7·58 inches at Pedong. On the other hand, Kurseong recorded 19·85 and Mangpū 16·85 inches; while there was a fall of 29·42 inches on the Tiger Hill tea estate, to the south-west of Ghum, and no less than 27·20 inches fell on the Happy Valley tea estate, close to Darjeeling. The result was a succession of very heavy landslips in the hills, one at Rangbī extending from an altitude of 7,000 feet to the bottom of the valley, where it had a width of 250 yards.

In the Kurseong subdivision only nine lives were lost, the storm not attaining its full fury till it approached close to Darjeeling. In the head-quarters subdivision outside the town of Darjeeling, the number of lives lost was 219, all natives. Deaths occurred more or less all over this area, but the loss of life was greatest at Pul Bazar, where 67 deaths were recorded. Here the bazar was situated on the banks of the Little Rangīt, which rose some 30 to 50 feet, carrying all before it; and it is presumed that numerous landslips dammed up its waters further up stream, with the result that huge masses of water were precipitated when the dams burst. At the same time, the Tīsta came down in flood of unprecedented height, and overflooding its banks caused wide-spread damage, especially at the Tīsta Bazar where nearly all the houses were swept away. The tea gardens suffered very heavily; the area of tea destroyed being nearly 2,000 acres, while the value of the property lost, *i.e.*, of tea, buildings, etc., amounted to over 10 lakhs of rupees. Large stretches of forest were swept away, the most serious loss being caused by the flood of the Bālasan river, which washed away three-quarters of the Bālasan forest into the valley. The roads were breached in all directions, the Cart Road being seriously damaged throughout its whole length from Darjeeling to Ghum, and also at many places between Ghum and Kurseong; while large portions of the Tīsta Valley road practically disappeared. The railway also was terribly damaged, the rails in several instances hanging in mid-air over gaping chasms; and great distress was caused by the stoppage of transport and the high price of food which resulted.

In Darjeeling itself the loss of life and property was appalling, 62 natives and 10 Europeans being killed, while the value of property destroyed amounted to lakhs of rupees: it is noteworthy that of the 72 deaths recorded, 45 occurred on the precipitous eastern side of the hill. Here, after a spell of fine weather, a thunder-storm burst at 1-30 P.M. on Saturday, the 23rd September, and there was a heavy downfall of rain till 4-30 P.M. A lull ensued till about 8 P.M., and then the cyclone burst in all its fury. The storm raged the whole of that night, and all the next day and night (Sunday) till about 3-30 A.M. Though small slips were noticed during Sunday, it was not till the evening of that day that the heavy rain began to have serious effect, and most of the worst catastrophes took place at midnight or in the early hours of the morning, when the tempest was at its highest. These disasters afforded opportunities for the display of bravery of the most conspicuous kind. Along the eastern side of the Mall, there was an almost continuous series of landslips, the majority of them starting from near the top of Observatory Hill. Many of the houses lay some distance down the hill-side, the approach to them was perilous in the extreme, and in one case the feat of reaching a house 100 feet below the Mall was described as equivalent to throwing oneself over the hill-side. There was the risk of all the rescuers being killed by fresh slips coming down from above There was the danger of being carried down the hill-side by the constant stream of mud, water, stones, etc.; of falling down the precipice in the numerous places where the paths were either blocked with trees, electric wires, and general debris or were entirely broken away; and of being buried under the houses which threatened at any moment to collapse and kill those who were digging out bodies below. The night was pitch dark; the rain fell in torrents; the danger was of unknown magnitude, and of an unprecedented kind, to face which required unusual courage. This was not wanting, and many lives which would otherwise have inevitably been lost were happily saved.

CAUSES OF THE LAND-SLIPS.

As regards the causes of these landslips, the following extract may be quoted from the report of the Special Committee which was appointed by Government directly after the disaster to consider the possibility of preventing the occurrence of landslips in Darjeeling:--

Nature of the land-slips.

"The landslips which occurred in such great numbers in Darjeeling were entirely confined to the soil-cap – the class known to the Swiss Geologists as *Schuttstürze*. As usual, they probably commenced as soil-cap creeps (*Schuttrutschungen*), but the heavy rain which immediately preceded the slips was so very excessive and sudden that the whole series of phenomena from the preliminary creep to the final catastrophe followed one another too rapidly to permit diagnosis of the usual premonitory warnings which precede all landslips. In many instances the slips can be traced to a preliminary fall of earth on to a high-level road, thereby blocking a drain and turning a large stream of water down the unprotected hill-face. In these instances, however, the slips in the lower part of the hill partake of the nature of a very rapid scour, reaching the foot as a river of mud and boulders.

Conditions for the development of landslips.

"Although it is true that the cause of the slips can be immediately traced to the heavy rain of September 24th—25th, it must not be forgotten—and this is a lesson of permanent value to all hill stations—that the necessary facilities have been in the course of gradual development for many years. The soil-cap is the direct product of the atmospheric decomposition of the rocks, and through the action of percolating meteoric waters, is in process of continual growth at the superficial expense of the latter. The removal of soil from the surface by the mechanical action of running water, and the simultaneous addition of decomposition products below the sub-soil by the chemical activity of percolating water are not concurrently compensatory in slopes covered with vegetation: there is a balance in favour of the latter process which is periodically restored by slips from the surface. The formation of a soil-cap does not in itself contribute to an increase in the surface slope, but its removal from the foot of a slope by streams, which there acquire a greater erosive activity through increased velocity and volume, increases the average slope of the hill-side. Such active undermining of the foot of a hill may be seen in a marked degree at the junction of two *jhorās*, as for instance, below the spur separating the two first branches of the Kāg *jhorā*, or where a river like the Rangnu cuts a line parallel to the lower contours of the hill. The increase in the angle of a slope by the undermining action of a river is supplemented by the slow process of creep which proceeds step-fashion in every inclined soil-cap with the regular succession of wet and dry seasons. The expansion which follows the saturation of a soil-cap during each monsoon naturally takes place in the direction of least resistance, which is down the hill-side. The desiccation which follows in the succeeding dry season merely makes a pause in this movement—not a return to original conditions. Such a ratchet and paul kind of creep downwards and outwards proceeds until the conditions of stability are exceeded, and a landslip occurs to restore equilibrium. It is delusive, therefore, to suppose that because a slope has apparently withstood the action of twenty monsoons, the absence of accident is an index to its stability and an insurance against danger in the twenty-first wet season. On the contrary, in a steep earth slope unprotected by artificial means. every monsoon brings it nearer to the inevitable landslip. and the more perfect appreciation of this fact by the authorities in our hill stations will help to guard against the constant, but fatal, tendency there always appears to be to repose confidence in a slope which has not by chance been the scene of an accident for a generation. The limits of stability having been established for slopes in different materials under various degrees of saturation, the rules so determined should not be superseded by the necessarily limited experience of any local authority.

"The angles at which free landslips are possible on earth slopes capping the undisturbed massive rock are not confined within areas where the rocks have been crushed by earth movements. Marked settlement, for instance, on slopes with an average inclination of 32° has been noticed between the two branches of the Kāg *jhorā*, where the subsidence has resulted in the production of serious cracks behind and above Manor Lodge and through Blossom Grove, and again on the eastern side of the ridge, where the W.N.W.–E.S.E. fault which determined the position of the Kāg *jhorā* has cut through the hill. In each of these cases the crushed condition of the rock has permitted the free percolation of water, which, issuing as springs at lower levels, has been enabled, through the loose texture of the fractured rocks, to carry away large quantities of material in suspension, instead of acting in the usual and much slower fashion by chemical solution. In these cases, besides the occurrence of free slips at lower levels, where the foot of the slope has been undermined, the higher levels of the slopes have undergone serious subsidence with the production of numerous ramifying fissures through the surface. Such cases of local subsidence by removal of material from the deep portions of the soil are necessarily rare in Darjeeling. The rocks are very uniform in composition, and include no specially soluble formations whose removal would result in destructive local subsidence. Practically, therefore, the only landslips which we have to deal with in this affected area are free falls of earth from highly inclined slopes. Local subsidences.

"The immediate cause of the numerous and destructive landslips of September 25th can be satisfactorily traced to the excessive rainfall of that and the previous day. Up to the 24th the monsoon rainfall had been some 17 inches in excess of the average, and although there had been a partial cessation during the previous week, there was no break in the rains sufficient to Cyclone and heavy rain of September 24th-25th.

permit of any appreciable drying of the soil before the unprecedented fall of the 24th and 25th. During the 24 hours ending at 8 o'clock on the morning of the 24th, 5·30 inches fell, followed by 19·40 inches before 4 A.M of the following day. Of the latter amount 14·32 inches fell between 4 P.M. on the 24th and 4 A.M. on the 25th, making an average of over an inch an hour for a stretch of 12 hours. The previous meteorological reports for Darjeeling show no parallel for this excessive precipitation, and falling as it did upon slopes already saturated by an unusually heavy monsoon, it may be with little doubt regarded as the immediate cause of the great damage done in the district. The heavy rain was an accompaniment of a severe cyclone which was first noticed by the Meteorological Department on the 21st as developing to the south-east of False Point in the Bay of Bengal, and from there moved gradually northwards until, on the morning of the 24th, the centre of the depression was near Rāmpur Boāliā, giving rise to a heavy rainfall all over the Province.

Severity of the winter of 1898-99.

"Residents in the station have called attention to the unusual severity of the previous winter; but the minimum temperatures recorded at St. Paul's School on the Jalāpahār ridge and at St. Joseph's College, North Point, show that the low temperatures were not sufficiently below freezing point or continuously maintained to seriously affect the soil-cap by frost, and so prepare the slopes for the reception and further destructive action of rain. The mean air temperature for the day never fell below 32° on the Jalāpahār ridge, whilst the minima were generally less than 3 or 4 degrees below freezing point. With or without a fall of snow, these temperatures are sufficiently mild to preclude any chance of freezing beyond the merest superficial skin of soil.

Absence of evidence of earthquakes.

"All enquiries which have been made as to the occurrence of earthquakes on the night of the 24th-25th have led to negative results. Whilst local earth tremors may have followed, and resulted from, the heavy landslips, there is no evidence to show that an earthquake was the actual cause. The negative results recorded by Mr. John Milne in the Seismological Laboratory at Shide on the Isle of Wight, and by the Meteorological Reporter in the Alipore Observatory, show that no earthquake of considerable magnitude could have occurred on the 24th or 25th, and as this is in agreement with the local evidence, the possibility of an earthquake being the cause of the numerous slips can be safely left out of further consideration.

"Both the positive and the negative evidence thus point to the heavy rain on the night of the 24th and 25th as the cause of the damage done in Darjeeling, and as the most excessive fall occurred as the final stage of a severe cyclone, and at the end of the rainy season, when the ground was already saturated, it is sufficient also to account for the approximately simultaneous occurrence of the large number of landslips in and near the station. It is just possible that the great earthquake of 1897 sufficiently weakened certain portions of the soil-cap to permit slips in places which otherwise might have successfully resisted the action of this storm But the effects of the rain so completely overwhelmed the other contributory causes that the storm may be looked upon as practically the only cause of the disaster."

Immediately after the disaster, a Committee was appointed, consisting of engineers and residents of Darjeeling, to inspect and report on the condition of buildings, roads, and drains in Darjeeling, and to devise measures to prevent the occurrence of landslips in future or to minimize their consequences. The Committee, who were assisted by Mr. Holland of the Geological Survey Department, made a thorough and searching investigation, visiting every part of Darjeeling, examining the condition of every building, road or drain which had suffered material damage, and determining the cause of the injury in each case. In their report the Committee showed clearly that the landslips were confined to the soil-cap, and that there was no reason to entertain any apprehension as to the stability of the site of the station, which is founded on massive rock and is thus secure. They clearly traced the causes which led to the landslips in each case, such as defective drainage of sites, excessive lead of drains, imperfect or badly-constructed revetments, neglect to reduce or protect steep slopes, defective supervision of building sites, quarrying in unsafe localities, etc., and indicated the remedial measures which were required in order to ensure safety for the future. These involved legislation. The Municipal law in force in Darjeeling was the ordinary mofussil Municipal Act, which had been framed with reference to the requirements of towns in the plains, and was in many ways defective and inapplicable to the circumstances of a town situated on a steep hill-side. A Bill was accordingly introduced, based on the recommendations of the Committee, which passed into law as Bengal Act I of 1900.

REMEDIAL MEASURES.

This Act made very important additions to the powers of the local authorities. All roads and bridges were brought under their control; and power was taken to enforce any alterations in them necessary to secure the stability of any hill-side or bank, or any buildings situated on them. Authority was given to enforce the repair and, in extreme cases, the removal of any building which threatened the security of a hill-side or bank, and to compel the owners to protect the sites when insecure. Similar provisions were enacted in respect of all drains; and power was also taken to enforce the construction of revetments and retaining walls, the turfing of banks and the sloping of the hill-sides to the angle of safety, whenever any of these works might be necessary for the general safety. At the same time, a complete set of building regulations was provided, which included full powers to regulate the excavation and preparation of building sites, and to prohibit building on any site considered insecure by professional authority. This was one of the chief dangers to be guarded against, and was the direct cause of much of the loss of life and property in 1899.

As soon as the Act came into force, extensive works were set on foot for the protection of all dangerous localities, and were pushed on vigorously, in order that as many of them as possible might be complete before the rain of 1900 began. The cost was very heavy, but success was complete. The rainfall of the next season was as heavy as usual, 17 inches falling in three days in June; yet not a single revetment or retaining wall failed; and the station enjoyed entire immunity from damage.

CHAPTER IX.

RENTS, WAGES AND PRICES.

RENTS.

Kālimpong Government estate.

THE rents paid by the cultivators throughout the district vary according to the class of land held by them. In the tract east of the Tīsta they are based on a fourfold classification of the land, viz., cardamom lands; terraced rice lands (*pānikhet*); ordinary or unterraced lands (*sukhākhet*), including the fields growing maize, wheat and other dry crops, orchards and fallow of less than three years; and waste lands, *i. e.*, lands which have lain fallow for more than three years. There is no uniform rate for each of the last three classes, as the amount of the rent is determined in each case by the fertility and general advantages of the block in which the land is situated, such as situation, slope and aspect, the cost of cultivation, outturn and profit, etc. The rates charged for terraced rice fields accordingly vary from 8 annas to Re. 1-4 an acre, those for dry cultivation from 6 to 15 annas, and those for waste lands from 2 to 3 annas, the average for the three classes being 13½, 8 and 2½ annas an acre respectively. Cardamom being a special crop, which yields nothing for the first three years, the land on which it is grown pays no rent for that period, but after three years pays a rent of Rs. 10 an acre. Besides this, land in the Kālimpong bazar is charged for at the rate of Rs. 4 per 100 square feet.

The average rate of rent, excluding bazar and cardamom lands, is 9 annas 10 pies per acre in the old surveyed tract, *i. e.*, the portion included in the valleys of the Tīsta, Rillī, Mayrung and Rashet, which was surveyed in 1882. Here the rates have hitherto been kept low in order to attract cultivators to the land, but such a reason now exists no longer, as the cultivated area in this portion of the Kālimpong estate has already expanded to its natural limits. In the other portion of the estate, which had not been surveyed prior to 1902-03, a poll-tax was levied until that year, and the cultivators were thus able to some extent to change their cultivation every year. This wasteful system has now been replaced by a regular land settlement under which they are only entitled to hold certain fixed pieces of land, the average rent rate being 7 annas 2 pies per acre. Besides the rents paid for the land, all tenants are required by customary obligation to furnish provisions and coolies for Government purposes at the current market rate, and to give free labour for two days in the year on the upkeep of the village roads.

West Tīstā Khās Mahāls.

In the West Tīsta Khās Mahāls the lands have been divided, for purposes of assessment, into three classes according to the quality of the predominating soil. The soil classed as first class is loose black earth, which is fit for almost all crops and is especially suitable for Indian-corn, while soil of the second class is black earth mixed with red, and that of the third class is red earth, which is said to be suitable for *maruā* (*Eleusine Coracana*). The rates for these three classes of land are twelve, nine and six annas an acre respectively, but the assessment is virtually reduced to nine, six and three annas owing to the liberal allowance made for fallow lands. As in Kālimpong, the cultivator pays no rent for cardamom lands during the three years in which there is no outturn, but after that pays at the rate of Rs. 10 an acre.

The Tarai.

In the Tarai the two chief classes of land are those known as *rupit* and *faringati*, *rupit* being the low lands which grow only the *haimantik* or winter rice, or admit of another crop before transplantation, while *faringati* are the high lands on which jute, Indian corn and *rabi* crops are grown. Each of these two classes of land is further subdivided, for purposes of assessment, into three groups, the lands being classified according to the rents paid by the sub-tenants. For this purpose the *jot*, *i. e.*, the land held direct under Government by the tenant under a 20 years' lease, is taken as the unit, and it is assumed that the lands are all first class in a *jot* giving a rent of Rs. 16 to Rs. 25 a *hāl*—a *hāl* is the area one pair of bullocks can plough and is roughly equal to five acres—while in *jots* paying a rent of Rs. 10 to Rs. 15 per *hāl* the land is considered second class, and lands paying a lower rent per *hāl* are regarded as third class. The rents charged for these three classes are, in the case of *rupit* land, Rs. 2, Re. 1-8 and Re. 1-4, and in the case of *faringati* land 10, 8 and 4 annas respectively.

The average rent per acre paid by the *jotdārs* or tenants themselves on the whole cultivated area is only Re. 1-5; but, owing to the practice of sub-letting, the actual rent paid by the cultivators is far greater. The *thīkādārs*, or lessees who hold the land on a cash rent immediately under the *jotdārs*, pay on the average Rs. 3-6-8 per acre or nearly thrice as much as their landlords pay to Government; while the *dar-thīkādārs*, who are tenants-at-will holding the land under the *thīkādārs*, pay Rs. 3-15-3 per acre. Besides these, there are a class of men known as *adhiārs*, who cultivate the land on condition that they get half the produce. They receive from their landlords, whether *jotdārs*, *thīkādārs* or *dar-thīkādārs*, advances of paddy to enable them to tide over the time till the crop is reaped, and are also given bullocks, ploughs and other articles necessary for the cultivation of the land and the construction of their houses. They are, in fact, practically hired labourers paid in kind.

The rates of rent for tea lands vary considerably in different parts of the district. In Kālimpong and in a few cases in the hill territory near Darjeeling one rupee per acre is paid, while six annas an acre is the usual rate near Darjeeling and in the Terai; but in the latter tract Re. 1-8 an acre is paid for any land originally included in the holding of a *jotdār* which has been acquired by tea-planters.

A statement of the wages current during the ten years 1893-1902 will be found in the Statistical Appendix. The wages paid to the tea coolies, who form the great majority of the labouring population, are considerably higher than such labour would obtain in many plains districts. The average rate is Rs. 6 a month for men, Rs. 4-8 for women and Rs. 3 for children, but in addition to the actual sum paid as wages, they receive substantial extras in the shape of free housing, fuel, medical attendance, etc. Outside the towns and tea plantations, the wages of unskilled labourers have remained almost stationary for many years; but on the other hand in urban areas the wages are high, and the great demand for skilled labour which has sprung up has greatly increased the rates paid to artisans. In 1871 masons received from Rs. 10 to Rs. 14 a month and carpenters from Rs. 12 to Rs. 18, whereas a mason can now make Rs. 15 to Rs. 30 a month according to his skill, a common carpenter Rs. 15 to Rs. 22, and a skilled carpenter Rs. 30 to Rs. 60 a month. In the towns the wages of all classes of labourers, artisans and domestic servants have risen considerably. The blacksmith and silversmith charge a higher rate for their workmanship; the shoe-maker and tailor have raised their tariff; and there is a similar tendency among domestic servants. Even the Bhotiā dandywala in private service generally earns Rs. 11 a month, or Rs. 10 if he is given a godown. If he plies for public hire, the rate is fixed by a definite tariff, which he is bound to observe or render himself liable to the penalties laid down by the Porters and Dandywalas Act. WAGES.

The price of food-grains has risen enormously during the last thirty years. In 1871 the ordinary price of the cheap rice PRICES.

eaten by coolies was Re. 1-4 a maund in the Tarai and Rs. 2-4 a maund in the hills, while the other staple food of the people, Indian corn, was sold at Re. 1-8 a maund. The price of these articles of food is now 50 per cent. as great; but fortunately the rise in the price of food has been to some extent counterbalanced by the general improvement in the material condition of the people which has taken place, the cultivators being benefited by the greater value they obtain for the produce of their fields, and the labourers by the high wages which they can obtain. In the Tarai, moreover, the poorest classes, the *adhiārs*, who correspond with the agricultural labourers elsewhere, are always paid in kind. They merely find the labour, and in return for this receive a half-share of the produce. Consequently they have been directly benefited by the rise in the price of food-grains; and this system also protects them in years of scarcity, for whatever the fluctuations in the market may be, their wage remains the same.

As regards articles other than the staple food grains of the native population, prices are extraordinarily high, owing to the fact that they have to be imported by the railway, which charges very high rates. For this reason, Darjeeling station enjoys the unenviable reputation of being one of the most expensive places in India for a European to live in.

MATERIAL CONDITION OF THE PEOPLE.

The district has for many years past made rapid strides in prosperity, and as a rule the people are very well off. In the hills they are blessed with an abundant rainfall and are immune from famine; there is a great demand for labour; wages are higher than in any other part of Bengal; and, generally speaking, there is no district in the Province where the people are so well-to-do. The best proof of this prosperity is perhaps afforded by the phenomenal increase of the population, which has partly

The Hills.

been caused by the natural growth which is one of the most obvious signs of general well-being, and partly by immigration from Nepāl, Bhutān and Tibet—a movement which indicates the opinion of the people themselves as to the relative advantages of two rival systems. They live in a stimulating and bracing climate under a Government the protection of which they or their ancestors have voluntarily sought; their industry is remunerative; and they are generally happy and contented, if only they are free from the iron fetters of the money-lender and unspoilt by contact with the shrewd and grasping immigrants from the plains. Owing to the difficulties of cultivation and the expense of transport, prices range higher than elsewhere in Bengal, but there is a great demand for labour of all kinds, and the greater expense of living is thus more than balanced by the high wages which can be earned.

Partial failures of the maize crop, which furnishes the staple food of the people, occasionally occur, but the peasants have resources unknown to the cultivators in the plains. Not only can they always obtain labour in the tea gardens and towns, where the demand is greater than the supply, but when the crops are poor, the forests provide ample food for those who like to make use of them. Even in ordinary years the Lepchās eat freely of the jungle products from choice and not of necessity—there are said to be more than 100 different kinds of forest fruit, fungi, etc., eaten by them—while in times of scarcity all classes draw upon this reserve of food. Poverty in the ordinary sense of the word is practically unknown, owing to the large amount of land still available, the constant demand for labour and the abundant facilities for petty trade. Any one who is in real want can find ample work, and among the local inhabitants beggars are practically unknown. The tea-garden coolies, in particular, can earn very good wages, and, provided they do not spend them in drink or waste them in extravagance, make large savings. A very good proof of the large amount they can save may be seen in the gold and silver jewelry and the strings of rupees with which their women folk are often loaded, and in the way in which they buy up land in the Government estates at prices which the older cultivators cannot or will not pay.

In the Tarai also the people may be regarded as in comfortable circumstances, so far as an actual sufficiency of the necessaries of life and a comparative degree of luxury are concerned. The tract is free from all fear of famine, the competition for cultivators protects the peasantry from ill-treatment, and they undoubtedly enjoy greater material prosperity at a smaller expenditure of labour than their fellows in other parts of Bengal. The ryots have always enjoyed a full measure of tenant right, and their holdings are sufficiently large to yield a comfortable subsistence. Indeed, they are often too large to be cultivated by one family, their size having been regulated by the facility of clearing jungle attaching to a system of extensive grants to men of some capital. The result is that, where the land has not passed to the money-lender, the *jotdār*, is as a rule, a substantial cultivator with a very considerable amount of capital. The system has, it is true, caused the growth of a class of under-ryots who are little better than labourers; but even this class is, comparatively speaking, well off. The number of labourers is small, and the wages are sufficient to attract immigrants from distant places. The poorest The Tarai.

person is thus sure of receiving food, not only as long as he can work, but also in temporary sickness, while there is a hope of work being obtained from him later.

But the ease with which subsistence is obtained in this part of the district is scarcely a matter for congratulation when the cause is considered. All the inhabitants suffer more or less from malaria, and the majority have in consequence some constitutional disease. The population has attained the stationary stage, and there is a slight tendency for it to decrease, in spite of the relief afforded by immigration. It is this that prevents wages being kept down and holdings being unduly subdivided under the influence of competition. Although, therefore, the inhabitants of this unhealthy region are pecuniarily well off, it must be admitted that the chronic misery caused by malaria deprives them of many of the pleasures of life, and that they are, on the whole, a depressed and unhappy race, whose dismal outlook on life is a great contrast to the cheerful, contented spirit of the hill people.

INDEBTEDNESS.

In spite of the general prosperity which prevails, the indebtedness of the people is one of the most serious economic problems of the district. This is due to the reckless manner in which they contract debts, owing to the customary obligation to incur heavy expenditure on marriage funeral and other ceremonies, to their love of drinking—they drink far more than the inhabitants of any other part of Bengal—to their propensity to gambling, to their simple delight in display of all kinds, which leads to an extravagant outlay on dress, ornaments and jewelry, and finally, to their improvident habits. When in want of money, they at once turn for a loan to the money-lender, the Mārwāri, or, as he is called locally, the Kanyā or Kayā. The big Marwāri merchants are, as a rule, honest in their mercantile transactions, able men of business who are strictly honourable in ordinary trade. But the principal is not the man with whom the borrower deals. The wealthy Kayā owns a number of shops, each of which is in charge of a *gumāshta* or agent who has orders to make a good profit. These *gumāshtas* are Mārwāri clerks working for Rs. 30 to Rs. 50 a month, they are promoted according to the profit they can show, and in some cases they receive a commission on the receipts. The result is that they follow the old adage "Make money, honestly if you can, but make it;" and they do this by charging exorbitant rates of interest, and by practices which are not too scrupulous. This is all the easier, because they have practically a monopoly of the money-lending business.

The money-lenders.

The ignorant hill people are recklessly willing to sign bonds at high rates of interest for ready money, interest accumulates on interest, and the unhappy debtor rarely manages to clear off his crushing debt. The money-lenders are always eager to accommodate a fresh victim; and are willing to allow him to become involved fully up to the limit of what they think he can be made to pay. Till that limit is reached, they are quite content to allow the principal to remain unpaid, while they gather in the interest month by month, and the unhappy debtor goes on blindly paying his monthly instalments. The reason for this complaisance on the part of the money-lender will be readily understood from two typical cases. In one a poor old woman was compelled by hunger to borrow 3 seers of rice, and only succeeded in getting a loan of that amount on condition that she repaid 30 seers of maize in two months time; in another a peasant, who had been driven by scarcity to borrow 10½ rupees worth of rice, had managed to pay off Rs. 76 in five years and still owed Rs. 140.

The difference between the original loan and the sum finally claimed is due not only to the exorbitant interest charged, but also to the ingenious devices of the usurer. To take a common instance, a *sardār* who wants to borrow Rs. 100, and agrees to pay at the rate of one anna in the rupee per mensem, has first to sign a note of hand for Rs. 106-4, *i.e.*, Rs. 100 plus a month's interest. At the end of the first month he owes Rs. 112-14-3, compound interest being charged on the original Rs. 106-4, and at the end of 12 months he has actually been charged compound interest at 6¼ per cent. for 13 months, and owes Rs. 219-14-9. The money-lender then gets him to sign a fresh note of hand for this sum as if it was a fresh loan, and in this note he states that he has received Rs. 219-14-9 and agrees to pay at the same rate of interest as before. Sometimes the old note of hand is destroyed, the result being that if a suit is instituted, the Court has no means of knowing what was the original sum lent, or it is surreptitiously kept by the Kayā and produced later for the purposes of the suit. In the latter case, his dishonesty is rewarded, as he makes it appear as if the borrower had signed notes of hand year after year for ever increasing amounts, and there is nothing to show that the larger amounts entered in the various notes really include the smaller sums for which the preceding notes were signed. Methods of usury.

Rates of interest vary according to the nature of the security given, but whatever the security they are invariably high. For loans on personal security, a common rate of interest is what is termed *ānā sud*, *i.e.*, the debtor has to pay one anna a month for each rupee lent, in other words, excluding compound interest, he Rates of interest.

pays 75 per cent. per annum Another usual rate is five rupees per hundred a month, in which simple interest alone would amount to 60 per cent. per annum. As already stated, however, compound interest is charged; and if the interest has not been paid off at the end of a year, the amount due is added to the principal, and a new note of hand taken for the increased amount, interest being charged on the total at the same rate as before.

The cultivators. Happily the cultivators in the Government estates are more or less protected against such extortion. As the representative of Government and the direct landlord, the Deputy Commissioner is in a strong position and can settle promptly and equitably a great many questions of a civil nature There is moreover an excellent rule that none but actual cultivators may hold leases for land outside the bazars of Kālimpong and Pedong This salutary provision has prevented the peasants from getting into the clutches of the professional usurers, who are not only unable to get hold of the land, but are also restricted in their money-lending operations. The tenants, consequently, borrow from one another, and the amount of indebtedness is comparatively small. This is particularly the case with the Bhotiās, owing, to some extent, to a curious custom observed by them. When a Bhotiā dies, his son or some other relation sometimes deposits in the Buddhist monastery a sum of money which is lent out for a year to another Bhotiā or Lepchā, as the case may be. At the end of a year, the latter has to perform certain ceremonies for the soul of the deceased, give a small present to the Lāma and pay back the principal. If he cannot do so, the debt is continued for another year, but in any case there is no interest to pay, but only the customary present to the Lāma. Even in this favoured tract, however, the money-lender has obtained a footing owing to the facilities afforded to them by the cultivation of cardamom. The fruit is grown solely for export and is therefore disposed of to Mārwāri traders, who use this opportunity to get the cultivators into their debt, by making advantageous bargains with them for the price of the unreaped crop and by paying out advances on the cultivation. The result is that, though large profits are often made on the crop, many of the simple Lepchās who grow it are miserably poor.

The tea gardens. Outside this tract the evil is very widespread, especially among contractors and tea-garden employés. Nearly all the contractors are financed by the money-lender. As soon as he receives a contract, the contractor obtains an advance on the strength of his agreement, he does not pay it off till a long time afterwards, and when he does his profit is swallowed up in the interest he has to pay. On the tea gardens the state of affairs is even worse. The people incur expenditure recklessly, they look to their *sardār* for money, and he gets it from the money-lender. The hold of the latter over the labour force is so great that it is credibly stated that in some gardens, paying Rs. 2,000 or Rs. 3,000 a month as wages to their employés, the Kayā regularly attends on pay day, takes the whole amount from the *sardārs* and then doles it out to the latter weekly, after deducting the interest due on the loans he has made. The general state of affairs may be illustrated by the experience of one of the leading planters of the district, who writes—"One *sardār* told me that he had never owed less than Rs. 2,000 for the last 20 years, and he said 'I believe that the day I get out of debt, I should die.' He meant that the environment under such circumstances would be so peculiar and abnormal that his constitution would never stand the awful and stupendous change. Three years ago I lent my people here about Rs. 4,000 at 6 per cent. by redeeming their *hāth-chittās* (notes of hand) for that amount from the Kayās. I cut the proportion of principal and interest weekly, and they now owe me only about Rs. 700. When they have paid up that, I shall have to give them more money as they cannot sleep, cannot eat, and cannot live without owing a good round sum to some one. Their argument is 'If somebody will lend me Rs. 100, why should I not take advantage of my splendid credit'. Any other course seems absolute lunacy!" These remarks at least show how blindly and willingly the Nepali cooly incurs debt, and the evil is aggravated by the extraordinary readiness with which his relatives will back his bills.

Statistics of indebtedness.

In these circumstances, it is not altogether surprising that the suits brought against debtors constitute a large portion of the civil work of the district; and a few bare statistics will show how prevalent the evil is and how exorbitant a rate of interest is charged. In the five years ending in 1902 no less than 808 cases were brought in the Civil Courts for debts due, the interest claimed ranging from Rs. 2-8 a month for each Rs. 100 lent, or 30 per cent. per annum, to Rs. 25 a month or 300 per cent. per annum. During this period 97 suits were instituted in which interest was claimed at 30 per cent.; 458 cases occurred in which interest at 36 to 37½ per cent. was charged; in 93 cases it was claimed at 48 per cent., in 38 cases at 60 per cent., in 113 cases at 108 per cent., in 3 cases at 150 per cent. and in one case at 300 per cent. When the records of the Civil Courts disclose such a state of affairs, it may easily be imagined how many cases of a similar nature occur which never come before the law courts. The majority are simply due to the fact that the hill people are extremely simple, as well as thriftless and extravagant, and are easily led by present necessity to agree to whatever is demanded in the shape of interest without thought for the future.

The evil is not confined to the hills, but is also prevalent in the Tarai. Here the inhabitants are a different race, but they largely consist of the indigenous Rājbansis, and of aboriginal tribes, such as the Oraons, Santāls and Mundās, who come to the district as tea-garden coolies and sometimes settle down as cultivators; they also include a large number of Nepālis. In this tract it was found in 1903 that out of 834 *jots*, with an annual demand of revenue amounting to a lakh of rupees, no less than 272 *jots* with an annual demand of Rs. 23,000, or nearly one-third of the total number, were owned by professional money-lenders. Altogether 150 *jots* had passed into their hands in the preceding 5 years, 139 being taken over from the local agriculturists and the remainder from other money-lenders, who held them previous to the transfer. Enquiries instituted more recently show that by the end of 1904 over 40 per cent. of the total number of *jots* had passed into the hands of money-lenders, and that even in the 30 years' grants as many as 12 per cent of the grants were owned by them. The difference is due to the fact that many of the latter belong to European tea-planters and intelligent native gentlemen, who are not likely to be cheated by the rapacious money-lender in the same way as the *jotdārs*, who are mostly illiterate farmers. This passing away of the Tarai *jots* from the hands of the local agriculturists is undoubtedly due to the ignorance of the people and the high rate of interest demanded by the usurer. The usual rate of interest until recent years was 75 per cent. per annum; this rate is still prevalent to some extent; and in the majority of cases the rate is not less than 36 per cent. per annum. All payments made by a debtor are credited as interest until all the sums due are paid up, so that the *jotdār*, like the cooly and cultivator of the hills, once in the clutch of the money-lender, does not escape till he has lost his property. He then usually stays on as an under-tenant of the new *jotdār*, but occasionally moves on to some other place and reclaims new land.

Remedial measures It is urged by some that the proximity of the frontier and the ease with which a debtor can escape into Nepāl without paying his debts, and the risk consequently incurred by the money-lender has necessarily kept the rate of interest in this district higher than elsewhere. This is, to a certain extent, true in the case of the tea-garden coolies who have no ties, but the Mārwāri generally takes good care not to lend to any one who has no prospects or can offer no security of any kind, and it is certainly not true of the ordinary cultivator. He cannot take his house or fields across the frontier; and where there is no house or other immovable property, the money-lenders take ornaments, jewelry and other valuables as a pledge, so that they can seldom be entire losers even if the debtors abscond. The reason for the high rate of interest charged seems rather to be that the number of money-lenders is few; that there are no banks which the cultivators and coolies can use, except the few agricultural banks recently started in the Kālimpong Government estate, where they are least needed; and that the Mārwāris, who are either connected with each other or form a close ring, have practically a monopoly of the usury. These Mārwāris form a very small minority; and in these circumstances it is small matter for wonder that many consider it necessary to take special measures to protect the simplicity of the Nepāli, who, though reckless, is very honest, from the rapacity of the professional money-lender. Proposals have, accordingly, been put forward to take legislative action by which interest should not be levied at a rate exceeding 24 per cent. per annum, whatever agreement the ignorant cooly or cultivator may have entered into, and to lay down that the total interest should not be allowed to exceed one-fourth of a one year's debt, and should not in any case exceed the whole amount of the debt.

Whether effect will be given to such proposals it is for the future to decide. But, in the meantime, much might be done by the establishment of agricultural banks or co-operative credit societies; and there also appears to be a very simple expedient by which the evil might at least be mitigated in the case of the tea-garden coolies who form the majority of the population. At present they are paid their wages once a month, and this is undoubtedly the cause of much of their indebtedness, as they are apt to spend or squander the money they receive before the next pay-day comes. If the same system obtained in England, it is doubtful whether the working classes would, for all their intelligence, contrive to get on without credit in some form or other with its necessary concomitants of debts, loans and interest. The tendency to contract debts is infinitely greater in the case of Nepāli coolies, who are ignorant, stupid and helpless in the hands of the money-lender, who are spendthrifts by nature—the men gamble and the women love velvets—and who live up to the limit of what they can beg, borrow or get credit for.

If they were to be paid weekly instead of monthly, the small wages they would receive would be no temptation to the Kayās, who are loth to give loans to men who do not receive a good lump sum on pay-day. It seems probable, therefore, that such an arrangement and the establishment of agricultural banks would tend, in some measure, to minimize the mischief which is at present caused by the professional money-lender.

CHAPTER X.

OCCUPATIONS, MANUFACTURES AND TRADE.

THE great majority of the population are dependent on agriculture, either as cultivators tilling their own fields or as labourers employed on the tea gardens. At the census of 1901 no less than 78 per cent. of the inhabitants were returned as engaged in agricultural pursuits, the number of rent-payers being 55,000 and of tea garden coolies 64,000. Altogether 63 per cent. of those dependent on agriculture are actual workers,—an unusually high proportion—while the numbers of agricultural labourers and rent-receivers are unusually small, the former aggregating only 500 and the latter being less than 200. The industrial classes are relatively unimportant, only 7 per cent. of the total population being engaged in manual industries. Of these, 60 per cent. are workers, who are distributed over various occupations, such as tailors, carpenters, masons, wood-cutters, etc. The commercial and professional classes are still smaller and account for only 1 per cent. each of the population. OCCUPATIONS.

In the hills the absence of certain functional castes, and in particular of the sweeper and washerman castes, is very noticeable. In Nepāl the Gharti or slave caste perform the functions of washermen and sweepers, but as soon as they migrate to Darjeeling they refuse to follow such degrading occupations. The result is that all the washermen and sweepers are men from the plains, the municipal sweepers in Darjeeling town being recruited from the distant State of Alwār. There are several other occupations which the hill people have failed to take up, such as those of barbers and carpenters. In the villages one member of the household will usually shave the others, but in the towns the barber's work is done entirely by Behāris. Similarly, they have not yet taken to the work of carpenters; the villagers use roughly-hewn wood in their houses so far as they use it at all, and all real carpentering work is done by Chinamen and carpenters from the plains of India. This is more surprising as they have learnt brick-making and masonry, and these new industries are now common to all castes. Certain occupations are confined to specified functional castes, such as the Sakris or cobbler caste, who make the native shoes and sheaths for *kukris*, the Kumal Newārs or potters, who are the only makers of earthern pots, and the Kāmis and Bhama Newārs, who manufacture metal ware and perform blacksmith's work. They are all low castes; and these manufactures are confined to them and will not be taken up by members of other castes.

MANUFACTURES. The most important industry in the district is the manufacture of tea, of which an account has already been given in Chapter VI; and next in importance is the manufacture of cinchona.

Cinchona. The cultivation of cinchona in Darjeeling was commenced in 1861-62. It had for some time, in fact since 1835, been considered desirable to extend to the hills of the Bengal Presidency the experiment of cultivating cinchona which had succeeded so well in the Nilgiris, the plants or seeds having originally been brought to the Madras Presidency from Peru under the superintendence of Mr. Clements Markham. The first cinchona seeds received in Bengal were some sent by Sir J. Hooker, in 1861, to Dr. Anderson, Superintendent of the Calcutta Botanic Garden, who conducted all the cinchona experiments in Bengal until he left in 1869. In 1861 Government took up the matter in earnest, and deputed him to inspect the cinchona plantations in Java. He received every assistance and attention from the authorities there, and brought back with him a large number of healthy plants. A few were retained for experiments in Bengal; the rest he took to the nursery at Ootacamund, whence other plants were brought to Calcutta. Dr. Anderson suggested the establishment of a cinchona nursery in Darjeeling, as affording the greatest hope of success: the proposal was approved, and a site was selected near the summit of Senchal in the midst of dense forest. The situation on Senchal proved too severe for cinchona; so in April 1863 the plants were temporarily removed to a garden at Lebong, a warm, well-sheltered spur below Darjeeling, at a height of 6,000 feet above the sea. For a permanent plantation space was found, 12 miles south-east from Darjeeling, in the Rangjo valley at Rangbī, on the south-eastern slope of a long spur projecting from Senchal at an elevation between 1,300 and 4,000 feet above the sea.

History of the plantations.

Here the cultivation, on an extensive scale, of those species of cinchona which contain quinine and allied febrifuge alkaloids in their bark was begun in 1864. The plantation was started with one hundred plants each of Cinchona succirubra and

Cinchona officinalis, and two plants of Cinchona Calisaya, at an elevation of about 4,000 feet. The stock of plants rapidly increased, so that ten years after the inception of the undertaking, there were nearly three million trees in existence, mostly of Cinchona succirubra, and the original clearing on the slope of the Rangbī had been extended in a south-easterly direction to the Rishop and Mangpu ridges in the Rangjo valley, while new extensions, comprising in 1881 about 750 acres, had been opened at Labdāh on the northern and Sitong on the southern slope of the Rayeng valley. It was soon discovered that Cinchona officinalis, the species yielding crown or Loxa bark, did not thrive, so that its further propagation was discontinued. For about the first decade the majority of the trees on the plantation were Cinchona succirubra, the species which yields red bark, poor in quinine but rich in a mixture of febrifuge alkaloids allied to quinine. The remainder of the trees were mostly of Cinchona Calisaya, or Ledgeriana, as it is now called, the species yielding yellow bark, rich in quinine.

In 1868-70 proposals were submitted by Dr. Anderson for the manufacture at the Rangbī plantation of a cheap but powerful febrifuge, well suited for use in native hospitals and charitable dispensaries, by separating the cinchona alkaloids from the young cinchona bark. The purchase of machinery for the experiment was sanctioned; and a factory was established at Mangpu in connection with the Rangbī plantation. This factory was equipped with the simplest of appliances for the extraction, by a combined acid and alkali process, of the mixed alkaloids from the red-bark trees; and in 1874 just ten years after the opening of the Rangbī clearing, the manufacture of cinchona febrifuge was begun, the first year's working yielding about 50 pounds of febrifuge. For the next fourteen years, up to 1887, only cinchona febrifuge was manufactured; but just before 1880 Dr. King, who was then Superintendent, initiated the policy of converting the plantation from one in which red-bark trees, poor in quinine, preponderated, into one of quinine-yielding species. In pursuance of this policy, the yellow-bark quinine-yielding species *par excellence* (Cinchona Calisaya or Ledgeriana) was planted out in gradually increasing numbers, together with a quinine which appeared spontaneously on the plantation about this time, yielding a natural hybrid between Cinchona succirubra and Cinchona officinalis. This substitution was pushed on with such vigour that, whereas in 1880 there were 4,000,000 red-bark trees to 500,000 yellow-bark and hybrid trees together, in 1890 there were over 3,000,000 of yellow-bark and hybrid to 1½ million of red-bark trees, and in

Cinchona febrifuge.

1901 over 2 million yellow-bark and hybrid to 200,000 red-bark trees.

Quinine. The object of Government in maintaining these plantations was to supply the hospitals and the people with a cheap remedy for malarial fever, which, as already stated, consisted up to 1887 of the preparation of cinchona bark, called cinchona febrifuge, a whitish powder composed of the alkaloids existing in the bark. In that year the manufacture of sulphate of quinine was commenced in the Mangpu factory by a process of extraction by fusel oil elaborated by Mr. Wood, formerly Quinologist, and Mr. Gammie, the Deputy Superintendent of the Plantations. From 1887 onwards the factory has continued to produce, in addition to cinchona febrifuge, sulphate of quinine in yearly increasing quantities; and the factory is now being extended in order that it may turn out in the future a minimum of 20,000 pounds. The issue of sulphate of quinine in 1887-88 was about 250 lbs.; in 1900-01 over 11,000 lbs., and in 1905-06 the output was nearly 16,000 lbs. In 1892 was instituted the system of selling sulphate of quinine to the public through the post offices in small packets, containing 5 grains (subsequently increased to 7 grains) at the price of one pice per packet, so as to enable even the poorest native to purchase a dose of the drug. In 1892-93 475 lbs., in 1900-01 3,400 lbs., and in 1905-06 4,200 lbs. of sulphate of quinine were issued for this purpose.

Extension of cultivation. From 1880 onwards the land near the original plantations in the Rangjo and Rayeng valleys proved too small for the number of trees which were required to keep pace with the increasing demand for febrifuge and quinine. Accordingly, in 1883 the first outlying plantation of 300 acres was started in the Ranjung valley in Kālimpong. But the rainfall here was too heavy to allow cinchona to be successfully grown, and the plantation was exhausted and finally given up in 1893. The Nimbong plantation of about 500 acres, also situated in the same tract, was purchased in 1893 from a private company. No extensions were attempted there, but the trees standing on the plantation as purchased were gradually used up, till in 1896 the last was taken up and the plantation abandoned. In 1899 a fresh extension of about 900 acres—since extended still further to about 7,000 acres—was commenced in the Damsong forest block, situated about 10 miles north-east of Kālimpong, near the junction of the Rangpo and Tīsta rivers on the borders of Sikkim. In this new block, which is known as the Munsong Division, there are at present about 500 acres under Cinchona Ledgeriana, with about 1,200,000 plants.

In the present year (1906), therefore, the Government cinchona plantations comprise the following :—(1) the Rangjo valley block, consisting of the Rangbī and Mangpu Divisions, which together measure about 900 acres, containing nearly over 2 million plants, of which more than a million and a half are Cinchona (Calisaya) Ledgeriana, nearly half a million of hybrid, and the remainder of Cinchona succirubra; (2) the Rayeng valley block, consisting of the Sitong and Labdāh Divisions which together comprise an area of about 600 acres, with over 200,000 plants, more than half of which are Cinchona succirubra and hybrid, and the remainder Cinchona (Calisaya) Ledgeriana; and (3) the Rangpo valley block comprising the Munsong Division, the details of which are given above.*

Where land is occupied by heavy jungle, the forest is felled and burnt a year before planting operations are to be carried out. As soon as the succeeding rains are ended, the ground is cleared again of its light growth of jungle, trenched to a depth of 1½ to 2 feet, roots taken out and stones collected in low transverse walls across the steeper slopes to arrest wash of soil in wet weather. The land is then staked out, from 5 or 6 feet apart according to the species to be planted out; and pits are dug in which the soil is again loosened and clods broken, the work being completed to admit of planting being done during the early spring. **Methods of cultivation.**

In appearance cinchona seed is small and chaff-like, weighing 60,000 to 70,000 to an ounce. This is harvested during February and March, and at once sown in prepared beds, protected from the weather by thatched watertight lines, with sloping roofs constructed some 5 feet high in front and 2 feet high behind, and facing north to prevent sunshine drying up the seed beds. When half an inch high, the seedlings are replanted in other beds an inch apart; and later, when they have attained a height of 3 to 4 inches, they are again transplanted to other thatched nurseries similar to those prepared for seed beds, and near to the land to be planted out. By October the seedlings will have completed their first year's growth and be a foot in height. The thatched covering of the lines is then removed, and the seedlings hardened off by exposure to sun until February, March or April, when they are planted out in their permanent places in the land prepared for them. Once growing weather has set in, the young plantation for the first year is kept

* I am indebted to Captain A. T. Gage, I.M.S., Superintendent of Cinchona Cultivation, Bengal, for this account of the development of the cinchona plantations in Darjeeling.

clear by hand-weeding about the plants and by sickling the intervening spaces. From the second year onwards, weeds are kept down by repeated light hoeing and hand-weeding.

Barking operations are carried out equally throughout the whole year, and are first begun when the trees are three years of age. Thinning then becomes necessary wherever overcrowding exists, and individuals that show signs of unhealthiness are uprooted. Every year the whole plantation is thus gone over and trees removed where necessary. To collect bark, the trees are uprooted and divided into three parts—root, stem and branch. All the bark is completely scraped off with blunt knives, and the three kinds—root, stem and branch—dried in open-air sheds, or in a heated godown during wet weather. Each kind is stored separately, and they are then taken for the extraction of the alkaloids to the quinine factory.

Cinchona Ledgeriana is the species most extensively grown. A hybrid between Cinchona succirubra and Cinchona officinalis was formerly grown largely, but the cultivation of this is being replaced by another hybrid between C. succirubra and Ledgeriana raised at Mangpu five years ago, from which selected trees yielding a minimum equal to 5 per cent. quinine sulphate are being set aside for future plantations. This latter promises to be successful in that the trees combine the strong habit of C. succirubra with the high quinine-yielding quality of the less robust C. Ledgeriana, which will admit of unoccupied land on the estate that is unsuitable for C. Ledgeriana being planted.*

Manufacture. The dry cinchona bark is first mixed with slaked lime and ground to a fine powder. It is then moistened and tipped into vats containing dilute caustic soda solution, which is heated by a steam coil li[illegible]ng each vat and continually stirred by a mechanical arrangement. Oil is then run on to the homogeneous mixture, the stirring kept up for two hours, and afterwards the whole allowed to stand till the oil has again separated completely from the bark sludge, carrying with it most of the quinine in solution. The remainder of the quinine is extracted by repeating the stirring with a second quantity of oil. The oil layers are run off from the top of the exhausted bark, and in another vat are stirred with water and sulphuric acid, which extracts the quinine and leaves the oil ready to be used again on more bark. The excess of acid in the aqueous quinine solution is then neutralized, and the crude quinine sulphate separates out

* The account of cinchona cultivation has been contributed by Mr. R. Pantling, Assistant Superintendent of the Cinchona Plantation, British Sikkim.

as a crystalline powder. It is purified from the other cinchona alkaloids by recrystallization from water, and from colouring matter by the aid of precipitants. The liquors from which quinine sulphate has been obtained are still saturated with it, and also contain all the other alkaloids from the bark. They are mixed to give a product of definite composition, colouring matter is removed, and then all the alkaloids are precipitated together by addition of caustic soda. This mixture of alkaloids after washing, drying and powdering constitutes cinchona febrifuge.*

European industries. There are no large European industries in the district, with the exception of the manufacture of tea, which monopolizes most of the labour available. There is a brewery at Sonāda, called the Victoria Brewery, which is managed by a European firm with branches at Simla, Kasauli and other hill stations. The output varies from 75,000 to 100,000 gallons per annum, the beer manufactured, which is of excellent quality, being supplied in large quantities to the Army Commissariat Department for issue to the soldiers stationed at Darjeeling. There is also a small brewery attached to the St. Mary's Seminary at Kurseong, at which beer is brewed for the consumption of the inmates. At Tung, not far from Sonāda, there are some private engineering works, which are chiefly occupied in fitting up factories on tea gardens, and in the construction and repair of the machinery used in the manufacture of tea. Besides these, there are the railway workshops of the Darjeeling-Himalayan Railway at Tindhāria, where a daily average of over 100 operatives are employed in the repair and manufacture of railway rolling-stock.

Native industries. The population of the district being almost entirely agricultural, the manufacturing classes are mainly engaged in supplying the simple needs of a rural people; and the native industries are practically small village handicrafts, producing little for export. The most important of these industries is weaving, coarse cotton cloth being woven by all the aboriginal races. The favourite patterns have a white or red background striped with red and blue or yellow and blue. These Lepchā cloths, as they are called, are in some request among the residents of and visitors to Darjeeling; and, as worn by the Lepchās themselves, they make a picturesque dress. Attempts are now being made to develop a village weaving industry among the peasantry in the Kālimpong hills; here the Church of Scotland Mission has recently started classes to teach improved methods of weaving, and

* This account of the processes of manufacture has been supplied by Mr. G. E. Shaw, Assistant Quinologist to the Government of Bengal.

has also introduced crochet work and lace-work. Knitting has for many years past been the occupation of the spare moments of the hill women, and they can everywhere be seen spinning woollen thread, as they walk along, to be made up afterwards into shawls, stockings, etc. Taking advantage of this aptitude, the missionaries are introducing more highly skilled methods, and women show great adaptability in learning and wonderful dexterity in working delicate crochet or fine gossamer-like lace, which is equal to some of the best European work. Government has come to the aid of the infant industry with substantial grants, and it is hoped that in time it will develop into a large and popular home industry.

The other industries are of small importance. In the Tarai ordinary coarse cloth, gunny-bags, mustard oil and common pottery form the chief articles of manufacture. These articles are however only produced on a very small scale. In the hills the artisans, for the most part, merely manufacture the articles of every-day use which are required by the cultivators or tea garden coolies, such as the baskets, trays and ropes made by Sharpās, the axes, chisels, hammers, spades and other iron instruments turned out by the Nepalese blacksmiths, the Kāmis, and the blankets woven by the shepherd caste, the Gurungs. Other articles of local manufacture are *kukris* and their sheaths, shoes, earthen pots and bamboo mats.*

MINES. The mineral products of the district include coal, iron and copper, but none of them appear to repay the expense of exploitation. The coal is easily exposed, but of a peculiarly friable nature; and the character and inaccessibility of the mines prevent their development by European capital. Owing to its flakiness, the coal cannot be used in its natural state, and it must be artificially compacted before it can be used as fuel, either by coking or by conversion into briquettes. For some time a company endeavoured to work the Dāling coal-field below Nimbong in the Kālimpong subdivision, but all work was given up a few years ago. The coal is badly crushed, and the lie of the seams is so much disturbed that it is difficult to follow them. Besides this, the cost of transport through the hills was prohibitive, and the poor quality of the coal did not warrant expenditure on improved methods of transport. Iron ore varying from a strong ferruginous clay to an impure brown hematite is found at Lohārgarh to the south-west of the district some distance below Pankhābāri. It is of poor quality, and is not smelted even by the natives,

* See also the account of the development of the industries of the district given at the close of chapter II.

though there are vague traditions of its once having been worked and of hammers still to be seen on the hillside, so heavy that no one can lift them. A bed of iron ore exists at Samalbong about a mile east-south-east of Sīkbhar to the east of the Tīsta, but this is no longer worked. Mr. Mallet considered the deposit a valuable one, the ore being extremely rich, while the quantity appeared considerable, and the iron produced was of the best quality. Copper ores, chiefly copper pyrites, exist in rocks of the Dāling series at Rānīhāt to the west to the Mahānadi near the mouth of the Baffupāni, at Pashok, at a place 2 miles north-east of Kālimpong, at Mangpu, in a ravine near Samthar, and in the neighbourhood of the Chel river. No attempt has yet been made to exploit them by European methods. Concessions have been taken out, but they have proved unproductive, while the number of mines deserted by natives goes far to show that even they do not find copper smelting in the Darjeeling hills a very lucrative employment. Practically, stone is the only mineral extracted, and this is procurable all over the hills from the rocks nearest at hand, such as the Daling beds, which yield coarse slate, the harder Tertiary beds near the foot of the hills, and the common gneiss, which usually is easily split into conveniently sized pieces.*

The main trade of the district is with Calcutta, the chief exports being tea, jute, gunny-bags, wool, cardamom and maize, and the imports being composed of European piece-goods, cotton yarn, rice, kerosene-oil and salt. Rice is also imported from Dinājpur, and coal and coke from the Rānīganj coal-fields. From Nepāl food-grains, cotton piece-goods, manufactured wool and hides, potatoes, sheep, goats, cattle and poultry are imported in return for European piece-goods and cotton twist, salt, kerosene-oil, tobacco and food-grains. The trade with Sikkim is of much the same character, but smaller in volume. Most of the Bhutān trade passes through the district of Jalpaigurī, but a fair quantity of the silk fabrics manufactured by the Bhutanese is imported into this district. The trade with Tibet has for many years past been much hampered by the jealousy of the Tibetan authorities; but as a result of the Tibetan expedition, it has begun to show greater vitality. The exports to Tibet consist chiefly of cotton piece-goods of European manufacture, and the imports of wool. Besides these articles, cotton yarn, silver, copper, brass and iron sheets, rice, maize and tobacco TRADE

* For a further description of the mineral resources of the district, see "Geology of the Darjeeling District and the Western Duars," by F. R. Mallet, F. G. S., Memoirs, Geological Survey of India, Vol. XI, 1875, pp. 51-96.

are exported, and yaks' tails, musk, horses, mules, sheep, and blankets are imported. The quantity of Darjeeling tea exported to Tibet is still insignificant, largely owing to the fact that the planters have not yet developed the manufacture of the brick tea which suits the Tibetan palate. The frontier trade is registered at 8 stations, Malli Ghāt, Rangīt, New Rangīt and Singlā being the registering stations for goods exported to or imported from Sikkim, Sukiā-pokhri for Nepāl, Kanjilia for Nepāl and Sikkim, Laba for Bhutān, and Pedong for Tibet and Sikkim and partly also for Bhutān.

Trade centres. The chief trade centres are Darjeeling, Kurseong, Silīgurī and Kālimpong. The three places first mentioned are on the railway, which carries most of the trade in the western portion of the district, while transport in the east is carried on by means of pack-ponies and bullock carts plying along the Tīsta valley. Kālimpong is the entrepôt for the Tibetan trade, and it is here that the wool of the Tibetan graziers is brought by mules and pack-ponies over the Jelep-lā pass. It is then packed in bales, and despatched to the railway at Silīgurī on bullock carts, which bring back cotton piece-goods and other merchandise for export to Tibet. The Tibetan dealers make Kālimpong their head-quarters from October to May, and the trade is almost at a standstill during the rains, as wool, which is the chief article of commerce, deteriorates rapidly if brought to Kālimpong and the plains in the moist heat of the monsoon months. The Nepāl trade passes for the most part through Sukiā-pokhri near the frontier on the road to Tanglu; but Simāna-bastī, 3 miles off, is rapidly becoming a important centre for this branch of trade, and Pul Bazar on the Little Rangīt is another bazar attracting some of the Nepalese merchandise.

In the interior of the district trade is carried on by means of weekly markets held in the larger villages. In the hills the principal of these weekly markets are held at Pedong on the borders of the district on the Tibetan trade route, at Sombāri at the end of the Chel valley, and at Pul Bazar and Sukiāpokhri. In the Tarai important markets are held at Mātighara and Naksalbāri, rice and jute from Purnea and Nepāl being sold there in large quantities ; and other markets of less importance are held at Phānsidewa, Bāghdogrā, Kharibāri, Adhikāri, Garidhura and Pānighatā.

Fairs. The most important fair in the district is that known as the Kālimpong *melā*, which is held on the last Wednesday and Thursday of every November. This *mela* was started in 1891, and has steadily increased in importance. It attracts not only the surrounding villagers, but also strangers from Nepāl, Bhutan and

Sikkim, as well as a large number of Tibetan traders, who bring in ponies, silks and mules for sale, the latter being bought up by Government for transport. An agricultural exhibition is held at the same time, prizes being given for grain, poultry, mules, ponies and all kinds of live stock. It is an annual festival for the whole country side, the villagers flocking in to see the athletic sports, dancing and other amusements ; but it has a very practical side, as it is most useful in stimulating both the agriculture and trade of Darjeeling. Another fair has also been started recently in the Tarai at Bāghdogrā.

CHAPTER XI.

MEANS OF COMMUNICATION.

DEVELOPMENT OF COMMUNICATIONS.

THE roads and railways of the district are a creation of British rule. Prior to their occupation of the country, the only roads were rough narrow tracks leading through dense forests, and the only bridges were the cane bridges spanning the torrents. One of the first measures taken by the British was to establish communication between Darjeeling and the plains; and between 1839 and 1842 Lord Napier of Magdala, then a lieutenant in the Engineers, was deputed to construct a road from Silīgurī to Darjeeling. This road was laid out in the midst of thick forest and along the steep ridges for a length of 40 miles. "On whichever side we turn," writes an early observer, "the whole road from Pankhābāri, upwards of 30 miles, runs through a forest with an almost impenetrable under-wood for the whole way. To be sure, here and there a Lepchā clearing, and now and then a mountain slip, on the top of a hill cleared by fire, or other accident, is free from trees; but they are hardly distinguishable unless upon the very spot." This road, now known as the old Military Road, may still be seen winding its way by sharp ascents from Pankhābāri to Kurseong, and then mounting up Dow Hill and running along the crest of the ridge to Ghum.

As Darjeeling developed and the tea industry became firmly established, it was recognized that this road, which was impracticable for wheeled traffic, was inadequate to supply its wants, and the need of having a cart road was still further emphasized by the Sikkim Expedition of 1860, when stores had to be hurried up to the front, and it was found that it cost Rs. 2 to transport each maund from Pankhābāri to Darjeeling. Government accordingly sanctioned the construction of a cart road, which was to be 24 feet in breadth, while the steepest gradients were not to be more than 1 in 18 or 20, the general gradient being 3 in 100. Work was begun in 1860, when the jungle from Darjeeling to Kurseong was opened out, and the construction of the road was commenced in 1861. The length from Darjeeling to Kurseong was open for traffic in 1864, and the whole road was completed in 1869. In the meantime, the communications with the plains had been improved by the construction, at a cost of Rs. 14,68,000, of a road 126 miles long, from Kārāgola Ghāt, on the Ganges opposite Sāhibganj, to Silīgurī; and a road had also been driven through the Tarai to connect this with the hill cart road. The former proved a work of great magnitude owing to the heavily flooded country it crossed, the section between the Mahânadi and Purnea, which is only 23 miles long, catching the drainage of 50 miles of country and being intersected by six great rivers; while the construction of a road through the Tarai was almost as difficult owing to the scarcity of labour and the fever-infested jungle which had to be cleared away.

But until 1878 the only means of approaching Darjeeling was by a long and tedious journey along the plains of Bengal. After the year 1860, when the East Indian Railway had been extended up to Rājmahāl, the traveller could proceed by rail as far as Sāhibganj on the banks of the Ganges, but having crossed the river, he was forced to proceed by road to the foot of the hills past Purnea, Kishanganj, Titalyā and Silīgurī in some conveyance such as a *pālki*, a ramshackle native carriage, or a slow-moving bullock cart. The whole journey from Calcutta lasted 5 or 6 days;[1] its discomfort can better be imagined than described; and its expense was equally great, Sir Joseph Hooker paying Rs. 240 for his conveyances from Kārāgola Ghāt on the Ganges to the foot of the hills. In 1878 the Northern Bengal State Railway was opened for traffic as far as Jalpaigurī, and by the end of that year it had been extended to Silīgurī. Tongas still continued to be the only means of travelling from that station to Darjeeling; but in 1878-79 a Company was started for the purpose of constructing, maintaining and working a steam tramway between the terminus of the railway at Silīgurī and Darjeeling. Government gave substantial aid to the enterprise by permitting the line to be laid along the cart road, and by guaranteeing that the gross receipts of the Company should not be less than 2 lakhs a year. The construction of the line was commenced in 1879 and was pushed on with great rapidity. The line was open for traffic as far as Kurseong in 1880, and in 1881 it was carried as far as Darjeeling, the steam tramway being then dignified with the name of a railway and its designation changed to the Darjeeling-Himalayan Railway.

The construction and maintenance of roads in the district are matters of great difficulty; in the interior, owing to the mountainous nature of the country, its heavy rainfall and its liability to landslips; and in the Tarai, owing to the great floods which swell the streams and rivers debouching from the mountains. In ROAI the hills the soil is a micaceous shale easily eroded and liable to landslips, especially where the forest covering has been destroyed and the rain ploughs through the exposed surface. The cost of road construction is extremely high, owing to the physical features of the country. The roads have to be driven up hill and down dale, often along the edge of steep precipices, where the hill side must be dug away, the rocks blasted, and the hill streams controlled. Expensive stone walls are consequently necessary, breast walls being built above the road to prevent the hill-side falling upon it, and revetments being built below to prevent it sinking. The rock-strewn streams which drain the mountain slopes are another source of danger. Trickling runlets in the hot weather, they become swollen torrents in the rains; and training walls have to be built far up the mountain to prevent them washing away the road. The effects of excessive rainfall have also to be guarded against, in order to prevent the roads being scoured out. For this reason, they are generally built with an inward slope and with a drain against the hill-side, as otherwise the steep inclines would make them mere water-courses; and at intervals there are small ridges of stone to divert the water into the drains. In some cases, the roads have been built with an outward slope, but in the opinion of many engineers this is a mistake, as the water rushing over it is apt to scour and undermine the hill-side below the road.

There are few roads with a gradient easy enough to allow of cart traffic, and the majority are built with a gradient which only admits of the passage of ponies and pedestrians: in some places, in fact, the gradient is so steep that ordinary metalling will not rest, and the roads have to be paved with rough stone slabs. These, though troublesome enough to walk upon, are the only means of keeping the road passable during the monsoon months, and of preventing them being scoured out by heavy rain. The roads have more often than not to be laid out in sharp zig-zags or curved round the spurs of the mountains and into the deep ravines and gorges; and the result is that their length is out of all proportion to the actual distance as the crow flies, being often twice and sometimes thrice as great.

Bridges. The bridges spanning the numerous water-courses, streams and rivers of the district furnish materials for an interesting lesson in the art of bridge building, varying as they do from the simple but ingenious bridges of cane and bamboo devised by the hill people to the more elaborate suspension bridges of European science. One type of native bridge, which is still found in places, is the *julunga*, which is built entirely with bamboos and canes cut from the neighbouring forest, and is remarkable for its lightness and extreme simplicity of structure. The following is a description of a typical bridge of this kind, given by Sir Joseph Hooker:—

"Two parallel canes, on the same horizontal plane, were stretched across the stream; from them others hung in loops, and along the loops were laid one or two bamboo stems for flooring; cross pieces below this flooring, hung from the two upper canes, which they thus served to keep apart. The traveller grasps one of the canes in either hand, and walks along the loose bamboos laid on the swinging loops; the motion is great, and the rattling of the loose dry bamboos is neither a musical sound nor one calculated to inspire confidence; the whole structure seeming as if about to break down. With shoes it is not easy to walk; and even with bare feet it is often difficult, there being frequently but one bamboo, which, if the fastening is loose, tilts up, leaving the pedestrian suspended over the torrent by the slender canes. When properly and strongly made with good fastenings, and a floor of bamboos laid transversely, these bridges are easy to cross. The canes are procured from a species of *Calamus*; they are as thick as the finger, and 20 or 30 yards long, knotted together, and the other pieces are fastened to them by strips of the same plant. A Lepcha, carrying 140 pounds on his back, crosses without hesitation, slowly but steadily, and with perfect confidence."

The piers for bridges of this kind consist of trees or rocks on either bank, or of strong piles propped with great stones or boulders. Sometimes, where the span is small, a truss consisting of bamboos is placed on each side of the bridge, these trusses forming huge loops stretching from one bank to the other.

Another bridge of an indigenous pattern is the log bridge built on the cantilever principle. Huge logs project upwards from either bank, where they are kept in position by heavy rocks piled over them, other logs are lashed over the first, and each successive layer overlaps the other until the projecting beams come close to one another. A small platform is then laid over the small space dividing them, and the bridge is complete.

These native bridges have now been largely supplanted by the more substantial and durable bridges of the Public Works Department. If the span is under 100 feet, Howe truss or girder bridges are built, and for longer spans suspension bridges are built. The latter have the advantage of not being destroyed by the heavy rainfall and great humidity which rot wood, bamboo and cane; they are not so liable to be swept away by floods; and they are cheaper than stone or iron girder bridges. It is, indeed, impossible to construct the latter over the larger rivers, because piers cannot be sunk in the rocky beds down which the water sweeps with the rapidity of a torrent. Besides this, it is easy and cheaper to bring long cables up and down the hill-sides, and they can be carried round corners which a large girder could not pass. A typical example of the suspension bridge now becoming common in the district is that over the Tīsta on the way to Kālimpong, which has a span of 300 feet between abutments and is swung 20 feet above flood level.

Conveyances. Strong bullock carts equal to very rough work ply along the few roads where the gradient is not too severe for wheeled traffic, but these roads are few in number, and the majority are paths too narrow and steep for any carts. Pack-ponies or human carriers are consequently most generally used for transport. The coolies are capable of great feats of endurance. It is not uncommon for a tea-garden cooly to carry a tea chest weighing 110 to 130 lbs. for a distance of 5 or 6 miles up an ascent of 2,500 to 3,500 feet; and others again, who bring in merchandise from the frontier States, perform long journeys of many days duration, carrying heavy loads, 150 to 200 lbs. in weight, along high ridges, up and down steep mountains, and through hot valleys varying many thousand feet in elevation. In Darjeeling itself the commonest conveyances for those who do not ride are the luxurious rickshaw and the hill dandy. The latter is a long coffin-like reclining chair with one end resting on the shoulder of a bearer, and the other slung across a pole (*dandī*) which rests on the shoulders of two men behind. For long journeys four men are necessary, and then the dandy is supported on poles both before and behind.

PRINCIPAL ROADS.

The Darjeeling Cart road. The most important road in the district is the cart road winding through the hills from Silīgurī to Darjeeling. This road, which is nearly 49 miles long, with an average breadth of 25 feet and a ruling gradient of 1 in 31, is one of the best mountain roads in India. As already stated, it was begun in 1861, in order to replace the old Military Road built by Lord Napier of Magdala, which was too narrow and steep for carts. By the year 1865 the upper section from Darjeeling to Kurseong, a length of 19½ miles containing 300 bridges, had been completed, at a cost of 5¼ lakhs, and was open for traffic; but the lower section from Kurseong to Pankhābāri and the road from that place to Silīgurī, which was to link up the cart road with the Ganges-Darjeeling road, were not completed till many years afterwards. Unexpected difficulties occurred. The Bhutān war drew off many of the labourers and the carriage required to bring up stores and materials, sickness broke out among the coolies in the lower valleys, and at the lower levels progress was interfered with by malaria and the annoyance caused by swarms of insects. In the Tarai, where the virgin forest had to be cleared, there was the greatest difficulty in getting any labourers to work at all, and when eventually they were secured, they could only be induced to stay by the energy and example of the Engineer in charge, who pitched his tent on the edge of the forest and hourly encouraged the workmen by his presence and determination. In the hills unlooked-for engineering difficulties obstructed the work. The ground was rocky and precipitous, the amount of blasting was far greater than had ever been anticipated, and at times the supply of gunpowder ran out. The difficulty of the work will be sufficiently illustrated by the fact that at a distance of about 3 miles below Kurseong a cutting had to be made in the face of a stupendous crag, a solid mass of rock about 500 feet in height and of the same breadth, that from Pankhabāri to Kurseong over 300 bridges were required, and that a section of only 6 miles from the latter place cost 2½ lakhs of rupees. By 1869 the whole road had been completed, and communication with the plains had been established.

This magnificent mountain road ascends from the foot of the hills as far as Darjeeling, a distance of nearly 49 miles, with an easy gradient throughout and without a single fall; and it is not too much to say that it is one of the finest monuments in the Province of the engineering skill of the Public Works Department. It is in charge of that Department, but its maintenance and annual repairs are entrusted to the Railway Company, which is remunerated for the work done by it by a commission of 15 per cent. on the amount expended.

Next in importance comes the Tīsta Valley road, which is the highway for travellers and merchants going to Sikkim and Tibet. It passes along the bank of the Tīsta, for a distance of 34 miles, from Sivok to Rangpo on the borders of Sikkim, and is exceedingly picturesque, running beneath steep high banks wooded to the water's edge with the Tīsta rushing down below over its rock-strewn bed. This river, with its rapid stream often swollen by floods, is however a great source of danger, as it is continually cutting away its bank and underscouring the road; and it has accordingly been decided to abandon the present alignment, and to construct a high-level road 300 to 400 feet above it. The new road is now being built at a cost of 5 lakhs, and is expected to be ready at the end of 1908. The present road is maintained by the Public Works Department from Sivok to Tīsta Bridge (17½ miles); **Tīsta Valley road.**

and from the Tīsta Bridge to Rungpo it is kept up by the Sikkim State.

Roads to the Tīsta.

At the southern extremity of this road there is an extension leading from Sivok to Silīgurī, a distance of 12 miles; and at the Tīsta bridge is an important junction where the roads to Darjeeling and Kālimpong meet it. The latter branches off to Kālimpong, mounting over 3,000 feet in 8 miles, the ruling gradient (1 in 15) being so severe that carts are unable to take a heavier load than 8 maunds up it; but there is also a track used by pedestrians which is 4 miles shorter and proportionately steeper. This road is continued to Pedong and then passes into Sikkim and over the Jelep-lā pass into Tibet.

Of the two roads from Darjeeling to the Tīsta bridge, the best is that running along the Takdāh ridge from Ghum, which is 17 miles long and sinks from an elevation of over 7,000 feet to 700 feet above sea-level. For the first three miles it winds through the great Rangarun forest and is fairly level, but then it dips down rapidly till the 6th mile, after which it ceases to be a broad cart road and is only passable by cattle, ponies and foot-passengers. It descends by sharp zig-zags through some magnificent forest scenery to Pashok (3,300 feet), where a beautiful view of the meeting of the Tīsta and Rangīt can be seen. At this point the woods are succeeded by tea cultivation, but after a short distance it again passes into semi-tropical forests, which continue till the Tīsta is reached. The other road from Darjeeling is unmetalled, and has been broken by landslips. It descends rapidly past Lebong to the dāk bungalow at Badamtam, 4,000 feet below and 8 miles distant from Darjeeling, where there is a magnificent view of the Great Rangīt 2,000 feet below. The road then dips rapidly down until the Rangīt is reached, after which it branches off to the right along its southern bank as far as the junction with the Tīsta; it then follows the course of the latter river till the Tīsta bridge is reached.

The Nepāl frontier road.

Another important road is that leading from Ghum to Simāna-bastī on the north-west. This is a good cart road, 10 miles long, leading along the Ghum range through the dense forest clothing its southern slopes. It passes through Sukiāpokhri (7 miles from Ghum) and 3 miles further on terminates at Simāna-bastī on the Nepāl frontier. Here it gives place to a mountain track; passable only by ponies and pedestrians, which leads over the high ridges and deep valleys along the frontier as far as Phalūt and Chiabhanjan, 40 miles to the north of Simānabasti.

Other roads.

Other important roads are those from Kurseong to Mātigharā *viâ* Pankhābāri (11½ miles), from Tiribāna to Bāghdogrā

(6 miles), and from Naksalbāri to Garidhura (11 miles); and in the east of the district from Rikyisum *viâ* Mimglās and Gorubathān to Jangi Guard (26 miles).

ADMINISTRATION OF THE ROADS.

The maintenance and construction of the roads are carried out by several agencies. The most important roads are in charge of the Public Works Department; they aggregate 336 miles, of which 173 miles are metalled and 163 miles are unmetalled, and are maintained at an annual cost of 1½ lakh. Besides these, 49 district roads, with a total length of 297 miles, and 5 village roads with a length of 10 miles, are kept up by the District Road Committee, at an average annual expenditure of Rs. 48-8 per mile, from the District Road Fund derived from the road cess. There are two branch road funds subordinate to the District Road Fund, each administered by Committees, which receive allotments of money for the roads maintained by them. That at Kurseong is responsible for the roads in the lower hills, and the Tarai Road Fund Committee for those in the Silīgurī thāna; while the District Road Cess Commitee is in charge of the roads in the hills to the east of the Tista. The Committee has no engineering staff under it, and in the hills the construction and repair of all the roads is entrusted to the planters,—an arrangement which makes for efficiency and economy, as the planters, who are vitally interested in the proper maintenance of the roads in the neighbourhood of their tea gardens, are careful to see that they are kept in proper order and that the labourers whom they employ to repair them do their work satisfactorily and give good value for the wages they receive. In the Tarai, where the number of tea gardens is fewer, an overseer is employed; but here too the repairs of most of the roads are executed by the managers of tea gardens, though the work is also occasionally done by contract. Both in the hills and the Tarai the more important bridges under the Committee are repaired by the Public Works Department at the cost of the Road Fund.

In the tea gardens themselves the roads are kept up by the tea estates concerned, and in the reserved forests, which occupy over one-third of the total area of the district, they are maintained by the Forest Department. To the east of the Tīsta in the Kālimpong Government estate, all the roads outside the forests and tea gardens are kept up by the free labour of the ryots. By customary obligation each adult member of a cultivator's household has to supply two days' labour, without payment, on the roads in or near the block in which he has his cultivation; and the *mandal* or headman of the block is responsible for keeping them in proper order. Altogether 158 miles of roads are kept up by the free labour of the villagers, and the system works well. It is cheap and, if properly supervised, efficient; and it is preferred by the ryots to any increase in their rent.

RAILWAYS. With the exception of a few miles of the Eastern Bengal State Railway south of the terminus at Silīgurī, the only railway in the district is the Darjeeling-Himalayan Railway, which has been laid out along the Darjeeling Cart Road. Starting from Silīgurī (398 feet above sea-level) it ends at Darjeeling, where the station is situated at an elevation of 6,812 feet, and it has a total length of 51 miles. For the first $7\frac{1}{4}$ miles, as far as Suknā, the line runs over a gently sloping plain, and the gradient is only 1 in 281; but after that station is past, it begins to climb the foot hills, and, as it ascends the slopes of the Himālayas, the gradient is far steeper. The highest point on the line is Ghum (7,407 feet), and after that station it descends for some 4 miles to Darjeeling with an average gradient of 1 in $31\frac{1}{2}$ feet. The average gradient from Suknā is about 1 in 29, but there are many parts steeper for short lengths, such as 1 in 26, and, for a short distance between Ghum and Darjeeling, 1 in 23. In places, the difficulties of the ascent are overcome by a few simple though ingenious engineering devices, such as loops, where the line mounts on long spirals, and reverses or zig-zags, by which the train is shunted up gradually ascending inclines. As a rule, however, the line merely follows the cart road, though here and there it leaves it when a shorter and easier alignment has been found practicable. The gauge is only 2 feet wide, and the engines are small but powerful, weighing 14 tons and being capable of taking a train of 50 tons up an incline of 1 in 25. The carriages are trollies open on either side, or miniature closed carriages, small, narrow, and low-roofed, both being slung very close to the ground with wheels only $19\frac{1}{2}$ inches in diameter.

To the ordinary traveller this railway, winding in and out among the mountains, below great crags and over deep precipices, appears a wonderful piece of engineering work, and it has frequently been described as such. This, however, appears to be an exaggerated estimate, as the road on which it was constructed was ready-made; and for the most part, it was only necessary to lay the rails along it. It is true that skill was required in seeing that the curves and radii were suitable and not too sharp, but no tunnelling had to be done, and the main difficulties had been already overcome by the engineering skill of the Public Works Department, which made the cart road many years before. The railway, moreover, cannot compare for speed, comfort and cheapness with mountain railways in other parts of the world, though it was no doubt a creditable achievement in the days when it was built.

It must however be remembered that the construction of mountain railways was more or less of an innovation in India at that time, and cheapness of construction and the safety of the line were matters of the first consideration. These essentials were secured by utilizing the existing road, though the speed of transit is necessarily less than if a shorter alignment with expensive bridging and tunnelling had been adopted.

Great credit, moreover, is due to the railway staff for the continued safety of the line, which is liable to be breached by torrents and landslips, laid out as it is on mountain slopes composed of friable shale, easily eroded by the streams pouring down their sides, and apt to crumble away and hurl down huge boulders. The portion of the line passing by the water course known as Pagla Jhorā or the "mad torrent" is a typical example of the difficulties which have to be encountered in maintaining the stability of the line and in keeping up through traffic. This torrent is situated near Tindhāria, half way between Darjeeling and Silīgurī, on the outer flanks of the Mahaldiram range, against which the monsoon current strikes with full force. It is the chief outlet of the rainfall of this range, and the result is that, after heavy rain, it becomes a furious torrent, bringing down huge rocks and masses of earth. The traces of its destructive work can be seen in tier after tier of old roads and revetments below; and it is the perennial task of the railway engineers to endeavour to control it and ensure the safety of the line. This they do by building training works far up the hill; by constructing revetment walls below the road, in order to prevent it sinking, and toe walls above it to keep the hill-side from slipping down, and when these measures prove insufficient and the road collapses with the sinking hill, by constructing a new road on a higher level. Here, as in other parts of the hills, the engineers are successful in their task; and the traffic along the line is rarely interrupted, except when there is a great catastrophe, such as that caused by the cyclone in September 1899, when long stretches of the railway were completely destroyed by stupendous landslips. The difficulty and costliness of keeping the line in repair is one of the causes of the high rates prevailing on the railway; but this is not the only cause, and the absence of competition is the chief reason why the railway is so expensive. The dividends paid are extraordinarily good, and the result is that, though the railway has been largely instrumental in developing the resources of the district, it may reasonably be contended that the high rates charged for transport have to some extent hindered its further growth.

POSTAL COMMUNICATION.

There are altogether 305 miles of postal communication and 33 post offices in the district. No detailed record exists of the number of postal articles delivered annually, but from an enumeration made in August 1905 and February 1906 it appears that on the average, 5,650 articles of all classes, such as letters, post-cards, newspapers and parcels, are delivered daily in the district. The value of money-orders issued in the year 1905-06 was Rs. 16,45,000 and of those paid Rs. 10,83,800; while the total amount of savings bank deposits was Rs. 1,14,700. There are also 14 telegraph offices from which over 55,000 messages were issued in 1905-06; these offices are situated at Darjeeling, Bāghdogrā, Ghum, Jalāpahār, Kālimpong, Kurseong, Lebong, Nāgri Spur, Naksalbāri, Pānighatā, Pankhābāri, Rangli Rangliot, Silīgurī and Sonāda.

TRAVELLERS' BUNGALOWS.

There are dâk bungalows at Kurseong, Pankhābāri and Silīgurī; and bungalows at the following places are available for the accommodation of travellers who have obtained passes authorizing them to occupy them :—

	Places.		Distance, in miles, from Darjeeling.	Height, in feet, above sea level.
1.	Badamtam	...	7½	2,500
2.	Jorpokhri	...	12½	7,400
3.	Kālījhorā	...	32 (*viâ* Tīsta Bridge)	550
4.	Kālimpong	...	28 (*viâ* the Rangīt) 32 (*viâ* Pashok)	4,000
5.	Mirik	...	26	5,000
6.	Pashok	...	18½	3,300
7.	Pedong	...	43	4,750
8.	Phalut	...	50	11,811
9.	Rangarun	...	7½	5,700
10.	Rayeng	...	25	625
11.	Sandakphu	...	37	11,929
12.	Senchal	...	6	8,000
13.	Tanglu	...	23	10,074
14.	Tīsta Bridge	...	19 (*viâ* the Rangīt) 22 (*viâ* Pashok)	710

CHAPTER XII.

LAND REVENUE ADMINISTRATION.

In giving an account of the land revenue administration of Darjeeling, it will be convenient to deal separately with the four tracts into which the district may be divided, viz., the Government estates in the hills, those in the Tarai, the strip of land lying to the north-west of the Little Rangit, and the land granted for the cultivation of tea. Besides these four tracts, however, there are reserved forests occupying an area of 415 square miles, and lands reserved for cinchona cultivation, which extend over 59 square miles.

KALIMPONG GOVERNMENT ESTATE.

The Kālimpong Government estate, including the Government reserved forests and the few tea plantations interspersed in it, comprises an area of 401 square miles to the east of the Tīsta, where it is bounded on the north by Sikkim and Bhutān, on the south by the district of Jalpaigurī, and on the east by Bhutān. The reserved forests in this area comprise cover 200 square miles and the tea gardens 10 square miles, and the actual area occupied by the tract managed direct by Government as a Government estate is 178 square miles. This tract was taken from Bhutān after the war of 1864, at the conclusion of which it was formed into a subdivision called Dālingkot. It was then placed under the Deputy Commissioner of the Western Duārs, but was transferred to the Darjeeling district in 1866. When first annexed, the population consisted almost entirely of Bhotiās, who paid a poll-tax in lieu of land revenue. The amount of this tax was Rs. 640 in 1865, which sum was collected by *mandals* or village headmen, who seem to have held a similar position under the Bhutān Government. By the year 1882 the revenue from this poll-tax had increased to Rs. 11,800, owing to the influx of new settlers, particularly to the north-west of the Tīsta, Rillī, Mayrung and Rashet valleys. Cultivation was more settled, the old inhabitants had become familiarized with the use of the plough and other agricultural implements introduced by the Nepalese, and there was consequently a great advance in the general prosperity of the people. This portion of the estate was accordingly surveyed and settled in 1882, the ryots' holdings were measured, rent rates were fixed for the different blocks, and *pattas* containing a statement of their rights were issued to the ryots. This settlement was made for 10 years, and on its expiry another settlement for the same term of years was carried out.

The last settlement, which is also to run for 10 years, was effected between 1901 and 1903, and was extended to the old unsurveyed tract in which the poll-tax continued to be levied. That system was in many ways suited to a primitive people inhabiting an undeveloped tract. Under it an active cultivator could occupy a large area without having his tax increased, but as the tract developed it was felt to be no longer suitable. It was found that in one part near Laba, where the land is the worst in the estate, being cold, stony and steep, about one-third of the cultivators had left the place owing to the heaviness with which the tax pressed upon them. Moreover, as the hillmen are very prone to wander and move from place to place, the system, though temporarily advantageous to them, was very wasteful; and it was the object of Government to remove from the cultivators all possible temptations to roam, and to encourage them to settle permanently. It was, accordingly, decided that in the unsurveyed tract the old poll-tax, under which the peasantry could to a large extent change their cultivation, should be replaced by a land settlement like that already introduced in the surveyed tract, under which the rent entitled the tenants only to certain fixed piees of land. At the same time, the old rates of rent, which were uniform throughout each block, but which varied from block to block according to fertility and accessibility, were replaced by a more elaborate and equitable classification of the lands within each block. The result of this settlement was to increase the rent-roll, excluding the Kālimpong bazar, to Rs. 31,600 or 45 per cent. more than ten years previously.

System of management. Government is the direct proprietor of the estate. There are no private landlords or tenure-holders, and there is no one between Government and the ryot, who is, in most cases, the tiller of the soil. For purposes of administration, the estate is divided into 48 blocks, each of which is under the charge of a headman or *mandal*. The latter is subordinate to the Manager, who again is under the control of the Deputy Commissioner.

Mandals. The *mandal* is the direct representative of Government on the estate. It is his duty to collect the rent due from the ryots in his block and remit it to the manager, to report all transfers of land, to inform the police of any crimes which may occur, to see that the roads in his block are properly maintained, to supervise the construction of any new roads that may be required, to ensure that the ryots provide labour and supply provisions for Government purposes, and generally to act as an intermediary between Government and the ryots in regard to the preservation of order and the public peace, the settlement of new land, the protection of trees, the maintenance of survey and boundary marks, and the care of grazing grounds and waste-lands in general. In return for these duties, he receives 10 per cent. on all his collections except the rent for cardamom lands, he has a rent-free grant of land. and he is sometimes allowed free permits for grazing. Besides this, he is the acknowledged head of the community; he is the arbitrator in disputes, except those relating to matters such as marriage, divorce and inheritance, which are settled by caste *panchāyats*; while his influence in settling questions of felling trees and transfers of land, which so vitally affect the interests of the tenantry, renders him a man of considerable importance.

The ryots of the estate are composed of Lepchās, Bhotiās and Nepālis, the three hill races for whom all the agricultural land has been reserved. The rent of their holdings is fixed for 10 years, and in the event of their failing to pay, their property may be sold up. They are strictly forbidden to sublet their holdings, or transfer any part of the land to others, unless they have previously obtained the consent of the Deputy Commissioner—a most important condition, as it is the only means of preventing the enterprising Nepāli from buying out the Lepchā, the rich cultivator from securing the land of his poorer neighbours, and the money-lender from ousting the actual cultivator by paying off his debt. The ryot is also obliged to provide coolies and provisions, when they are required by Government, at the market rates; and each adult, male or female, is bound to give two days' labour free of charge for the maintenance of the roads on the estate. Any infringement of these conditions renders the ryot liable to have his lease cancelled. The average holding per ryot for the whole estate is 9·70 acres, and the average rent of each is Rs. 6-8. **Ryots.**

Three different kinds of sub-tenancies have sprung up, owing to the practice of subletting. In some cases the sub-tenant pays all the cost of cultivation, except the rent, and reaps all the profit, paying a money rent to the tenant. In these cases the tenant is called a *pakuriā*. In others, the tenant lets out the land for a fixed produce rent, the subtenant, who is known as a *kuthdār*, paying all the cost of cultivation. Lastly, the tenant engages another man to cultivate his holding, on condition that the latter is given the seed and receives half of the actual produce after deducting the value of the seed advanced to him. In this case the sub-tenant is called an *adhiār*. Besides these, there are a certain number of under-ryots **Under-ryots.**

holding land in the rent-free grants of the *mandals*, monasteries and hermitages.

Mortgages. In addition to these different kinds of sub-tenancies, here are two kinds of mortgages. The first is known as *masikatā*, and is a mortgage, by which the mortgagee is given possession of the land for a fixed term of years, on condition that at its expiry the land reverts to the mortgager. Under this mortgage, both principal and interest are liquidated, and it thus differs from the mortgage known as *biyāz*, in which only the interest is paid off by letting the mortgagee retain possession of the land.

WEST TISTA KHAS MAHALS. The other Government estates in the hills are situated to the west of the Tīsta in the tract annexed from Sikkim in 1850. They comprise 15,000 acres and extend from Mirik in the Bālāsan valley, close to the Nepāl frontier, as far as Lopchu and Mangwā in the Rangīt and Tīsta valleys. There are altogether 14 villages, viz., Takdāh, which is divided into two blocks, Bidur Singh's and Pemchada's Takdāh, Lopchu, Barbatiā and Rishi-hāt, all situated within 4 or 5 miles from Darjeeling; Tumsong near Bloomfield; Poomong, Humlingding and Mangwā, 2 miles distant from Rangli Rangliot; Pulongdong, Pagriangdong and Parmagiri at the same distance from Sukiā-pokhri; Dhajiā near Nāgri; and Mirik, which, as already mentioned, is close to the Nepāl boundary. These estates are under the charge of the manager of the Kālimpong Government estate, and the system of administration through *mandals* and the terms under which the ryots hold their lands are similar to those in vogue in that estate. The last settlement was effected in 1893-94 and expires in the current year (1906-07).

THE TARAI. The Tarai is the tract lying at the foot of the hills, which is bounded on the north by the mountains, on the south by the Purnea district, on the east by Jalpaiguri, and on the west by the State of Nepāl. It has a length from north to south of 18 miles, and a breadth from east to west of 16 miles; excluding the area covered by forest, it contains a total area of 230 square miles. This tract was acquired by conquest from Sikkim in 1850. After the annexation, it was at first decided by Government that the southern portion of the tract should be placed under the Collector of Purnea, and the northern portion attached to Darjeeling. But, apparently in consequence of the extreme dislike shown by the inhabitants of the southern Tarai to the transfer to Purnea, Government decided in the same year to attach the whole tract to Darjeeling. Before this was done, however, a settlement of the land revenue for 3 years had been effected by the Collector of Purnea with the Rājbansi and Musalmān inhabitants of the lower Tarai. Meanwhile, Dr. Campbell, the Superintendent of Darjeeling, had made a settlement for 3 years of the upper Tarai, which at the time seems to have been chiefly inhabited by Meches. Previous to the annexation, the revenue was derived from a *dāo* or hoe tax paid by the Meches and Dhimāls; from lands settled with the Bengali inhabitants of the lower Tarai; from dues paid for cattle sent from the adjoining districts of Bengal to graze during the early months of the year; from forest produce; from excise or spirits; from market dues; from fines; and from a tax on musicians. The revenue raised from the first two at least of these sources was collected by Bengali officers called *chaudhris*, who also exercised certain civil and criminal powers, and were apparently in all cases *jotdārs* or holders of large grants of land.

The cultivated portion of this tract was first regularly settled for 10 years in 1853, and on the expiry of this period the settlement was renewed from year to year till 1867, when a fresh settlement was made for 10 years. A third settlement for a similar period was made in 1879, its term being subsequently extended to 1895. Besides the area thus settled, over 28,000 acres were granted in the years 1865-67 on 30 years leases for tea cultivation under the former Waste Land rules. The last settlement was completed in 1895, the total revenue being assessed at Rs. 1,13,222. In the case of the *jots*, *i.e.*, the lands set aside for the cultivation of ordinary crops, this settlement was made for 20 years from 1897, while the lease of the tea grants was renewed for another term of 30 years.

Land tenures. There are two classes of tenants in the Tarai holding their lands direct from Government, the *jotdārs*, who hold leases of land for ordinary cultivation, and the tea-planters holding grants of land for tea cultivation. The rights of both these classes are heritable and transferable, and are governed by the terms of their leases, which have been granted on condition that after the expiry of the period for which they run, the holdings will be resettled with the lessees and grantees on such terms as may be determined by Government.

Jotdārs. The ordinary cultivators are divided according to the legal incidents of their holdings into tenants-in-chief (*jotdārs*), sub-tenants (*thīkādārs* and *dar-thīkādārs*) and labourers paid in kind (*adhiārs*). The *jots* vary in size from 2 acres to 1,190 acres, and there is a corresponding difference between the position of those who hold them. Some are pleaders or men of business, who have purchased the holdings as a speculation or investment, some are poor ryots. The average size of a *jot* is little under 150 acres, and where the land has not passed into the hands of the money-lender, the ordinary *jotdār* is generally a substantial farmer representing the original reclaimer of the soil, and holding at an easy rent. Unfortunately, however, the unfettered right of transfer has been a fatal gift to the ignorant Tarai cultivator, and the result is that in a very great number of cases he has fallen into the hands of the money-lender and has lost his holding. In over 40 per cent. of the *jots* the money-lender is the real proprietor, and he is a mere annuitant, letting out the holding to others.

Thikādārs. The *thīkādārs* are lessees holding land on a cash rent immediately under the *jotdār*. They have no right of occupancy or of transfer by sale or gift. They are sometimes as rich as the ordinary *jotdārs*, many of them holding some land under one tenure, and some under another; but a large number, on the other hand, are very poor and dependent. The average size of their holdings is 9·6 acres. Lower down in the scale of sub-infeudation are the *dar-thīkādārs*, who hold their land under the *thīkādārs* and are mere tenants-at-will, the average size of the holding being 2·76 acres.

Adhiārs. At the bottom of the scale come the *adhiārs* or *prajās*. These are rather labourers paid in kind than tenants; they receive from their principal not only the use of the land, but the seed, the use of ploughs, and often food on account, giving him in return half the produce. The other half they retain, and hence they are called *adhiārs* or half-men. They form a distinct social class, being composed of men without capital, and often without skill, under the direction of a superior; but some of the actual tenants are in a position only a little less dependent, being immigrants or promoted *adhiārs*, without capital or credit, and quite unable to stand alone.

CHEBU LAMA'S GRANT. The tract known as Chebu Lāma's grant comprises an area of 49 square miles in the north-west of the district between the Nepāl and Sikkim boundaries and the Little Rangīt river. It forms part of the tract, measuring 115 square miles, which was annexed in 1850 in consequence of the treatment received by Dr. Campbell and Sir Joseph Hooker at the hands of the *Dīwān* of Sikkim. After the annexation, Chebu Lāma received a lease of the whole tract for the term of 3 years at an annual rental of Rs. 20, in recognition of the services he had rendered during the troubles with Sikkim. This lease was subsequently renewed, and in 1862 Chebu Lāma asked that, in consideration of his services, the land held by him should be granted in fee-simple or in perpetuity at a nominal rent. Government acceded to his request by making a grant of the land to him and his heirs for ever, in proprietary right, subject to the payment of land revenue to Government of Rs. 500 during his lifetime and of Rs. 1,000 after his death. Subsequently, on his death, this tract was leased out jointly to Rechuk Dīwān, the son of Chebu Lāma, Pharbu Dīwān, his brother, and Rājā Tenduk Palger, his nephew and adopted son. In 1882-83 part of the estate, containing the forests of the Singālilā range and covering 66 square miles, was sold to the Forest Department; and the remainder has been partitioned among the descendants of Chebu Lāma. The largest portion, containing 31 square miles, which is known as the Relling estate, belongs to the widow and daughters of Rechuk Dīwān, and has recently been placed in the charge of the Court of Wards, owing to the financial embarrassment of the proprietors. Another share of 15½ square miles, which is known as Rājā Tenduk Palger's estate, is owned by Kumār Polden, the eldest son of the Rājā, who has lately succeeded to the property on its release from the management of the Court of Wards. The last share which is very small, its area being only 2½ square miles, is held by one Sonam Sring, the adopted son of Tenduk Palger.

These are the only permanently-settled estates in the district; and they pay between them a revenue of Rs. 1,000 per annum, the demand being divided among the three estates in proportion to their area. They consist for the most part of mountainous country with small villages scattered on the spurs and ridges. They contain some patches of forest, but the land is mostly under ordinary cultivation.

TEA AND OTHER GRANTS.

The remainder of the district consists of the tract originally ceded by Sikkim in 1835 and of two strips of land annexed after the Sikkim war of 1850. The first tract is the old hill territory stretching from the Sikkim frontier to the foot of the hills below Pankhābāri, an irregular strip of land about 24 miles long and from 5 to 6 miles broad, containing an area of 138 square miles. The second tract contains 245 square miles, and consists of two strips of hill land, one lying to the west of the old hill territory and extending up to the Nepāl frontier, and the other lying to the east of that territory and extending to the Tīsta.

In this portion of the district there is a bewildering variety of land tenures. For some short time after the cession of the old Darjeeling territory in 1835, there appears to have been but little demand for land; and the applications which were made were dealt with by the Superintendent at his discretion, but in 1838 a large number of applications for land for building sites led to the issue by Government of a set of rules for the grant of land. These rules provided that the conditions of any grant made previously by the Superintendent should be binding on Government, but that, in future, land should be given only as follows:—(1) land suited for building locations, for which purpose a space of 200 yards broad, on either side of the principal road from Kurseong to Darjeeling, was specially reserved. (2) cleared spaces of undefined size, to be reserved for bazars, at Pankhābāri, Kurseong, Mahaldiram and Darjeeling; and (3) land not required for either of the above purposes, but available for farming leases.

Building locations.

The building leases, according to the original rules, were to be in perpetuity, subject to a rent of Rs. 50 for a full location, which was ordinarily to be 100 yards square, but this limitation was disregarded from the first. The Court of Directors modified these terms by limiting the period of lease to 99 years; but before their orders were received, 65 full locations and 10 half locations had been allotted in perpetuity. After the receipt of these orders, 76 full locations, 45 half locations, and 24 quarter locations were granted for terms of 99 years. The holders of these location leases were allowed the option, under the rules of 1859, of commuting them into fee-simple, at the rate of 20 years' purchase of their annual rent; and a large number of locations were commuted under those rules.

Bazar lands.

As regards the second class of land, a space of ground was reserved for the Darjeeling bazar, which is now under the management of the Municipality. There are also some Government reserves at Kurseong and Pankhābāri; but the land intended for a bazar at Kurseong, together with the Mahaldiram lands, has been alienated in course of time. The area set aside for the Darjeeling bazar was 42 acres. The Municipality has the management of it, and has expended large sums on buildings, from which a great part of its revenue is derived. Besides the bazar land, 47 acres were reserved for native settlers, the rents of which are also received by the Municipality.

Farming leases.

Lands for agricultural purposes were, under the rules, to have been leased in lots of not less than 10 acres, for a term of 30 years. If the land was uncleared, it was to be held rent-free for 5 years, and to pay Rs. 2 per acre for the rest of the term. It does not appear however that a single grant of farming land was made under these rules, which were rescinded in 1851, when it was directed that, in future, such leases should be granted at rates to be fixed by the Board of Revenue. Between 1851 and 1869, several farming lots were granted for terms varying from 10 to 99 years, and at various rates, some of which were apparently commuted afterwards into fee-simple tenures.

It appears from a report submitted by Dr. Campbell, Superintendent of Darjeeling, in 1850, that previous to that year he had not found it practicable to take any revenue from the aboriginal inhabitants of the old Darjeeling territory. In 1850, however, he tried the experiment of settling defined tracts with the headmen of the communities living within such tracts, for periods of 3 to 5 years; and he claimed that he had in this way settled the whole of the territory. In 1853 all these leases were renewed for 5 years. When they were first granted, there was no demand for land at Darjeeling, and the chief object was to attract native cultivators to settle under the lease-holders, who were bound not to take a higher rent than Rs. 2 from each household of cultivators. But soon after the second leases were granted in 1853, more serious attempts than had previously been made were begun to introduce the cultivation of tea in these hills. In 1858 Major James, then acting as Superintendent, granted several leases for periods of 50 years, for 5 years of which the rent was to be a fixed sum irrespective of area, and for the remainder of the term at the rate of 8 annas per acre. In the applications, these lands were described as required for farming purposes, but it was understood that they were intended for tea. The leases of 1853 expired in 1858; some of them were summarily renewed for 1859, and then nothing seems to have been done until 1860.

Meanwhile, a new set of rules for the grant of waste lands in the Darjeeling territory was issued in 1859. Their most important provisions were that grants of waste land should be put up to auction at an upset price of Rs 10 per acre; that the sale at such auction should convey a free-hold title; that existing lease-hold grants might be commuted to free-hold under the rules, at the option of the grantee; and that building locations might be commuted at the rate of 20 years' purchase of the annual rent. Between the introduction of these rules in 1859 and thier abrogation on the introduction of the fee-simple rules in 1862, over 9,000 acres of land were sold in the hills by public auction at an average rate of about Rs. 12 per acre. The provisions under which the lands were put up to auction were, however, much disliked; attempts were constantly made to evade them, some of which were successful; and commutation deeds were given to people who had no claims beyond having purchased the interest of a former lessee in an indefinite lease, the term of which had expired. *Waste Land Rules of 1859.*

In 1861 Lord Canning issued a minute regarding the sale of waste lands in fee-simple, which laid down three main principles on which grants of waste lands were to be made in future. These were, firstly, that such lands should be granted in perpetuity as a heritable and transferable property, subject to no enhancement *Fee-simple.*

of land revenue; secondly, that all prospective land revenue would be redeemable at the grantee's option by a payment in full when the grant was made, or a sum might be paid as earnest at the rate of 10 per cent., the remainder being paid later; and thirdly, that there should be no condition obliging the grantee to cultivate or clear any specific portion within any specific time. The minimum price for the fee-simple was fixed at Rs. 2-8 per acre, so that by paying 10 per cent. of this, or 4 annas per acre, a title was obtained. This minute was followed up by the issue in 1862 of fee-simple rules for the sale of land by auction to the highest bidder above a fixed upset price; and subsequently a large quantity of land was commuted to freehold by special orders which allowed the commutation of all farming leases given prior to the introduction of the fee-simple rules.

Waste Land rules of 1864. In 1864 farming leases for 30 years were introduced by Government under what are known as the old Waste Land rules. The lands granted under these rules were given free of rent for the first 5 years, and then were subject to an annual rate of rent amounting to 6 annas per acre on the whole area. At the expiry of 30 years, the lessee had a right of re-settlement at an increased rate not exceeding 50 per cent. of what the annual rental would be if cultivated with ordinary crops, in other words, at half the rates paid for land cultivated with the native crops of the district.

Rules of 1882. In 1882 another set of rules was issued for the lease of waste land for tea cultivation. Under these rules a preliminary lease was given for five years, the land being rent-free for the first year, and then paying a rental of 3 annas an acre for the second year and an additional 3 annas for each successive year until it amounted to 12 annas an acre. After that, the lease was renewable at a rate of Re. 1 per acre, if the conditions of the preliminary lease had been fulfilled.

Existing rules. These rules have now been replaced by the Waste Land rules which were promulgated in 1898, and which are still in force. Under these rules a preliminary lease of waste land is given on condition that the applicant shows that he has *bona fide* sufficient capital to open out the grant. Each lot has to be compact, and must ordinarily not contain more than 1,500 acres, though larger grants may be made on special grounds. A preliminary lease is granted for 5 years, the land being rent-free for the first year and paying a progressive rent thereafter. The rights conveyed are heritable and transferable, subject to certain provisos, *e.g.*, that the whole lot is transferred, that the conditions requiring the clearance of the land have been duly observed, etc. The right of Government to minerals and quarries, and to payment for valuable trees on the grant, and the right of the public to fisheries, and a right of way along the banks of navigable streams are reserved. Provision is also made for public access to springs of water on the land leased, when it is necessary to ensure a supply to persons residing in the vicinity, for the construction and maintenance of proper boundary marks, for the presence of the lessee himself or of a resident manager on the grant, for the supply of reports regarding the births and deaths of residents on the area granted, and for information as to the progress and outturn of cultivation. Government also reserves to itself the right to acquire any land in the grant which may be required for public purposes free of cost, except by a proportionate reduction in the rent and by the payment of the value of any improvements in the land taken up. Where the provisions of the Chaukīdāri Act are in force in the tea garden, the lessee is bound to furnish accommodation for the residence of the *chaukīdār* and to see that he receives his pay punctually. Grantees can club or amalgamate their grants; but no isolated grant can be amalgamated with other grants lying more than 2 miles away from it. If, after inspection during the term of the preliminary lease, it is found that 15 per cent. of the total area has been brought under cultivation and actually bears tea plants, the lessee is entitled to a renewal for a term of 30 years, and to similar renewals in perpetuity, provided that Government may fix the rent on certain specified conditions on each renewal. Subject to this condition, the renewed lease is heritable and transferable, in whole or in part, with due sanction and after proper registry; and all the other conditions of the preliminary lease hold good. Failure to comply with any of the conditions renders the lessee liable to forfeiture of his lease, and failure to comply with the clearance conditions of his preliminary lease reduces him, if he is allowed to continue, to the status of a tenant-at-will, the term of grace being limited to 3 years, within which, if he clears 15 per cent. of the total area, he may obtain a renewed lease. Government reserves the right to exclude any particular area from the operation of the rules and to sell grants in that area by auction; and declared forest reserves, land having valuable timber in compact blocks, lands in which other rights exist, lands lying within 60 feet from the centre of any public road, and lands expressly exempted by Government are not to be granted.

Building site rules.

The grant of building sites belonging to Government were regulated by special rules laid down by Government in 1879.

These rules prescribed that the lease applied for should be sold by public auction, the rent should be at the rate of Rs. 50 per acre, the size of the location should be limited to 2 acres, and the lease should run for 99 years. These rules however have not been strictly followed, and the special orders of Government are obtained in giving out such grants, each case being decided according to its merits.

Special reserves. There are also certain small areas held for special purposes. Since 1866 a strip of land on either side of the Cart Road from the plains to Darjeeling has been set aside for road purposes, though in some places settlers have been allowed to build houses; a small revenue is derived from these persons, who are mere tenants-at-will. At Jalāpahār, Katāpahār and Lebong certain areas have been made over to the Military Department; an area of 116 acres below Darjeeling has been set apart for the jail; and 622 acres, originally intended for grazing grounds, have been retained by the Darjeeling Municipality.

Synopsis. A return prepared in 1903 gives the following synopsis of the land tenures of the district and of the area held under each: –

Revenue-paying tenures.

	No.	Area in acres.	Area in square miles.
1. Locations held in perpetuity under the building location rules of 1839	29	139	0·22
2. Locations for 99 years held under the above rules as modified by the Court of Directors in 1840	102	367	0·57
3. Farming leases under the rules of 1859 ...	11	1,558	2·42
4. Cultivation leases for 30 years under the rules of 1864	131	52,125	81·45
5. Chebu Lāma's grant as now held by his heirs ...	3	31,634	49·43
6. Tea-cultivation leases under the Waste Land rules of 1882 as modified up to date ...	8	6,370	9·95
7. Other recent leases from 10 to 50 years ...	52	10,399	16·25
8. Government Khās Mahāls under direct management	6	221,047	345·39
Total ...	342	323,639	505·68

Free-hold tenures.

	No.	Area in acres.	Area in square miles.
1. Locations commuted into fee-simple under rule 10 of the rules of 1859	79	525	0·82
2. Lands brought under rule 1 of the above rules	50	48,476	75·74
3. Land commuted under rule 9 of the above rules	98	30,176	47·15
4. Lands brought under fee-simple rules of 1862	24	11,152	17·44
Total ...	269	90,329	141·15

Lands held by Government.

		Area in acres.	Area in square miles.
1.	Lands held by the Forest Department for Government Forest Reserves*	278,400	435·00
2.	Lands held by the Military Department ...	406	0·63
3.	Lands held by Government for cinchona cultivation	37,702	58·91
4.	Lands held by the Jail Department ...	116	0·18
5.	Municipal grazing land	622	0·97
6.	Other lands (*i. e.*, waste lands, bazar lands, etc.) ...	13,746	21·48
	Total ...	330,992	517·17

* The area of the Forest Reserves has now increased to 445 square miles.

CHAPTER XIII.

GENERAL ADMINISTRATION.

ADMINISTRATIVE CHARGES AND STAFF. THE district of Darjeeling is a non-regulation district, *i.e.*, some of the general Regulations and Acts in force in other parts of Bengal have not been extended to it. It was formerly part of the Rājshāhi Division, but, in consequence of the territorial redistribution which is popularly known as the Partition of Bengal, it was transferred in October 1905 to the Bhāgalpur Division. For general administrative purposes, it is divided into two subdivisions, the head-quarters subdivision, the northern portion of the district, which extends to the frontier of Sikkim and Bhutān and includes all the country east of the Tīsta, and the Kurseong subdivision which comprises the lower hills and the Tarai. The former, which extends over an area of 726 square miles, is under the direct control of the Deputy Commissioner, who is assisted by a staff of Deputy and Sub-Deputy Collectors stationed at Darjeeling, while the tract east of the Tīsta known as Kālimpong, which is a large Government estate, is in charge of a Manager subordinate to the Deputy Commissioner. The Kurseong subdivision, extending over 438 square miles, is in charge of a Joint Magistrate, under whom is a Deputy Collector, stationed at Silīguri, who disposes of the criminal work of the Tarai and manages the Tarai Government estate. Formerly, the district was divided into two subdivisions, the head-quarters subdivision, with an area of 960 square miles, which included all the hills on both sides of the Tīsta, and the Tarai subdivision, with an area of 274 square miles, which included the whole country at the foot of the hills. The head-quarters of the latter subdivision were at Hānsquar (Hānskhāwā) near Phānsidewā from 1864 till 1881, but in that year they were transferred to Silīgurī, in consequence of the importance that place had acquired as the terminus of the Northern Bengal State Railway. In the meantime, Kurseong had begun to develop, and, accordingly, in 1891 it was made the head-quarters of a new subdivision, including both the Tarai and the lower hills west of the Tīsta.

THE DEPUTY COMMISSIONER. The Deputy Commissioner is the head of the local administration, and performs a number of duties which do not fall to the lot of a Collector in an ordinary regulation district. He exercises the powers of a Sub-Judge in disposing of appeals from the Munsifs of Kurseong and Silīgurī, but has no power to entertain civil suits of first instance. He is vested with the powers of a District Delegate, and in that capacity deals with uncontested applications for probate of wills and letters of administration; and he also disposes of intestate cases and of any applications to be declared insolvent which may be made over to him by the District Judge. A large portion of his work consists of the administration of various funds and local bodies. As Chairman of the District Road Fund Committee and of the Kurseong Branch Road Fund Committee, he is responsible for the maintenance of a large number of roads scattered over the hills and the Tarai; as Chairman of the Darjeeling Municipality, he controls the administration of the affairs of the station; and both as Deputy Commissioner and as Vice-President of the District Committee of Public Instruction, he is in close touch with the education of the people. He is also *ex-officio* Chairman of the Darjeeling Town Improvement Committee, of the Darjeeling Dispensary Committee, of the Natural History Museum and Water-works Committees appointed by Government, and of the Managing Committee of the Lowis Jubilee Sanitarium.

DARJEELING IMPROVEMENT FUND.

One of the most important functions performed by the Deputy Commissioner is the administration of the Darjeeling Improvement Fund. This is a fund pecular to the district, which is known officially as an excluded local fund. It was formed in 1868 in place of the old Location Fund, which derived its income from the quit-rents levied in the old hill territory, an area of 138 square miles streething from Pankhābāri to the northern boundary of the district, and spent it on conservancy and local improvements. The income of the fund is now Rs. 77,000, of which Rs. 50,000 are derived from the ground-rents and fees levied in the various *hāts* and bazars owned by Government. The remainder of the revenue comes from a number of sources, such as location rents, ferry tolls, dâk bungalow fees, rents from land leased out for cultivation, and the one-anna cess levied in the Tarai. This income is expended on a variety of objects, which would in regulation districts be provided for by the District Board. The fund maintains a number of dispensaries, meeting the cost of their upkeep and providing the pay of the Civil Hospital Assistants and staff, besides making considerable grants to other hospitals and dispensaries in the district. Large contributions are made to the primary education grant for expenditure on schools, and to the District Road Fund for expenditure on roads. Provision is made for the proper sanitation of villages and bazars, and for the maintenance of a water-supply at Kālimpong, Pedong and Sukiāpokhri. A large number of staging bungalows are maintained along the principal routes, ferries are kept up, grants are made to agricultural fairs and exhibitions, and the funds necessary for the upkeep of the Natural History Museum at Darjeeling are provided. Besides this, it provides the pay of a Veterinary Assistant, whose duty it is to deal with outbreaks of epidemic diseases among cattle in the interior, and to minimize the risk of such outbreaks by regular inspections.

Darjeeling Town Improvement Fund. A special grant is also placed annually at the disposal of a Committee formed for the improvement of the station of Darjeeling. From this grant the Lloyd Botanical Gardens and the Victoria Pleasaunce are maintained; and the grant is also utilized in beautifying the town by planting trees, shrubs and flowers on the banks and slopes, and generally in securing the improvement of the station.

Kurseong Improvement Fund. A similar grant, though smaller in amount, is also made for the improvement of the town of Kurseong, the sum allotted for this purpose being administered by a small Committee presided over by the Deputy Commissioner.

The one-anna cess. The income derived from what is known as the one-anna cess is spent entirely in the Tarai. This cess was first assessed at the settlement of the Tarai *jots* in 1879-80 with the object of effecting local improvements and of developing the Tarai. Before that the greatest difficulty was found in keeping up the schools, as the *jotdārs* were very reluctant to subscribe, and the people could not be prevailed upon to pay any fees. The same difficulty was found with dispensaries. Only one had been established, and it was scarcely possible to obtain sufficient funds even for this, although it was of the greatest benefit in the unhealthy Tarai, and was freely used by the people. To remedy this state of affairs, it was decided to levy a cess on the *jotdārs*, who pay a very light revenue, of one anna for each rupee of rental paid by them, and to spend the amount thus realized exclusively on schools, dispensaries and works of local improvement in the Tarai. The annual demand of the cess is a little over Rs. 6,000, and more than this sum is spent in the Tarai for local purposes; for, besides the grant for primary education, over Rs. 4,000 is contributed for dispensaries, the number of which has now increased to seven, and a grant of Rs. 4,000 per annum is made for the maintenance of roads.

REVENUE. The revenue of the district under the four main heads increased from Rs. 1,62,000 in 1880-81 (when the income-tax had not been imposed) to Rs. 3,79,000 in 1890-91 and to Rs. 5,45,000 in

1900-01. In 1905-06 it amounted to Rs. 5,78,000, of which Rs. 1,67,430 were derived from land revenue, Rs. 2,43,660 from excise and opium, Rs. 70,350 from income-tax, Rs. 51,540 from stamps, and Rs. 44,820 from cesses.

Land revenue. The demand of land revenue is less than in any other district in the Province, except Angul and the Khondmāls, amounting in 1905-06, to only Rs. 1,67,430. Of this sum more than three-fourths, or Rs. 1,39,860, is payable by the Government estates under the direct management of Government. There are only three permanently-settled estates with a demand of Rs. 1,000 per annum, these being the three estates into which, as described in the preceding chapter, Chebu Lāma's grant has been partitioned; and the remainder of the land revenue, or Rs. 26,570, is payable by 158 temporarily-settled estates, which are all Government lands leased out for definite periods for tea cultivation.

Excise. By far the largest portion of the revenue of the district is derived from excise, *i.e.*, from spirituous liquor, beer, opium and hemp drugs; the income from this source in 1905-06 was Rs. 2,43,660, or 50 per cent. greater than that accruing from land revenue. Nearly two-thirds of this amount, or Rs. 1,59,260, was realized from the sale of the country spirits prepared by distillation from molasses and the flower of the *mahuā* tree (*Bassia latifolia*). There was only one country-spirit shop to every 43 square miles, serving an average population of 9,230, but over 60,000 proof gallons were issued from these shops during the year. Besides the strong distilled liquor, the hill people drink a great quantity of the mild native beer known as *pachwai*, which is brewed from rice or the *kodo* millet (*Paspalum scrobiculatum*). This is the national drink of the people, and they are therefore allowed to brew it up to a limit of 20 seers without payment of any fee. If they wish to brew a larger quantity, they can do so on payment of a fee of Rs. 2 each, which entitles them to brew as much as they please for domestic consumption; while payment of a fee of 8 annas enables them to brew any quantity for festive occasions. More than 7,000 licenses for the home-brewing of *pachwai* have been issued, besides 31 licenses for its sale; and the fees paid amount to Rs. 41,420. Besides this, nearly Rs. 18,000 are realized from the license fees paid on imported liquor, which is mostly consumed by the European residents, and Rs. 5,000 from the duty and license fees charged for the beer manufactured at the Victoria Brewery at Sonāda and at the St. Mary's Seminary at Kurseong. The beer manufactured at the brewery is supplied to the European soldiers stationed at Darjeeling, or sold to the public. The consumption of *gānja*, *i.e.*, the dried flowering tops of the cultivated female hemp plant (*Cannabis sativa*) and the resinous exudation on them, is not very great, bringing in only Rs. 13,000; while the receipts from the license fees and duty on opium amount to only Rs. 6,680. The drug is used mainly by the Chinese residents in the district and by immigrants from the plains, and is not in much request among the hill people.

The latter are, however, very heavy drinkers, and are unfortunately very much addicted to spirits. The mildly stimulating rice beer called *pachwai*, which is almost as much a food as a beverage, is drunk to a large extent, especially by Bhotiās, but the Nepalese prefer the strong country spirit produced by distillation. This is drunk in large quantities by the labourers, who exercise no self-restraint when once on a drinking bout; while the better class of natives who can afford such a luxury are acquiring a taste for European spirits and indulge in them on festive and ceremonial occasions. The incidence of the excise revenue in the district is now over one rupee a year for every man, woman and child; it is twice as great as in any other district in Bengal; and it is more than seven times as great as in the Province as a whole. The amount spent on liquor has been steadily on the increase in recent years, rising from Rs. 1,82,000 in 1894-95 to Rs. 2,23,000 in 1899-1900 and to Rs. 2,85,000 in 1904-05, *i.e.*, by over 55 per cent. in the decade. This great expansion of revenue is ascribed to the increase of the wealth of the district and to the growth of the population; but it is noticeable that during the decade ending in 1901 the increase in the receipts from country spirits was more than ten times as great as the increase in the population, and it appears therefore that the expansion of the excise revenue is largely due to the steady growth of drinking habits.

Income-tax. From the Statistical Appendix it will be observed that in 1901-02 the income-tax yielded altogether Rs. 61,000 payable by 1,452 assessees, of whom 582 paying Rs. 6,640 had incomes of Rs. 500 to Rs. 1,000. At that time the minimum income assessable was Rs. 500, but this was raised in 1903 to Rs. 1,000 per annum, and the number of assessees fell to 835. The growing wealth of the district has, however, prevented this measure having any effect on the total receipts of the tax; for though the number of assessees has decreased to 794, the collections amounted in 1905-06 to Rs. 70,350.

Stamps. The total revenue derived from stamps is small, amounting in 1905-06 to only Rs. 51,540, of which Rs. 33,140 were realized from the sale of judicial stamps and Rs. 18,400 from non-judicial stamps. Altogether Rs. 31,350, or more than three-fifths of the total stamp revenue, were due to the demand for Court-fee stamps, and Rs. 13,560 were received from the sale of impressed stamps. This is not a very elastic source of income, as the people are not litigious, and it shows little signs of increase. The average receipts for the five years ending in 1899-1900 were Rs. 48,550, and during the next quinquennium they were only Rs. 46,260; but as shown above they increased last year by nearly Rs. 3,000.

Cesses.

The road and public works cesses are as usual levied at the maximum rate of one anna per rupee. A revaluation was effected in 1904-05, and the total demand now amounts to Rs. 44,820, of which Rs. 26,470 are payable by 210 revenue-paying estates and Rs. 18,350 by 172 revenue-free estates. The number of estates assessed is thus 382, and these are held by 1,915 recorded share-holders; while the number of tenures assessed is 1,272, there being altogether 1,317 recorded shareholders of these tenures. The demand is now nearly Rs. 9,000 more than it was 10 years ago; and the gross rental on which the cess is levied is Rs. 6,94,420 as compared with Rs. 5,29,160 at the first valuation.

Registration.

There are three offices for the registration of assurances, viz., Darjeeling, Kurseong and Silīgurī. The Deputy Commissioner is *ex-officio* Registrar of the district, and the office of Sub-Registrar is filled at Darjeeling by the senior Deputy Magistrate, at Kurseong by the Subdivisional Officer, and at Silīgurī by the Deputy Magistrate in charge of the Tarai. The average number of documents registered annually at these three offices was 541 in the quinquennium 1895—99, and rose to 725 in the five years ending in 1904. From the marginal statement, giving the salient statistics for 1905, it will be seen that in that year the number was 681, of which more than half were registered at Silīgurī.

Name of office.	Documents registered.	Receipts.	Expenditure.
		Rs.	Rs.
Darjeeling ...	258	1,280	1,545
Kurseong	66	133	65
Silīgurī	357	572	474
Total ...	681	1,985	2,084

Administration of Justice.

The chief Civil and Criminal Court is that of the District and Sessions Judge of Purnea, whose head-quarters are at Purnea, but who occasionally visits Darjeeling. The district was formerly under the jurisdiction of the District and Sessions Judge of

Dinājpur, but since October 1905, when Darjeeling was transferred to the Bhāgalpur Division and Dinājpur was incorporated in the new Province of Eastern Bengal and Assam, it has been included in the jurisdiction of the Judge of Purnea.

Civil Justice.

At Darjeeling the senior Deputy Magistrate is vested with the powers of a Small Cause Court Judge up to Rs. 500, and of a Sub-Judge and Munsif within the head-quarters subdivision. At Kurseong the Subdivisional Officer has the powers of a Munsif and of a Small Cause Court Judge up to Rs. 50, and has authority to try rent suits; while the Deputy Magistrate at Silīgurī has the powers of a Munsif as regards civil suits instituted in the Tarai. Appeals from the decisions of the Munsifs of Kurseong and Silīgurī lie with the Deputy Commissioner; and the District Judge of Purnea hears those preferred against the orders of the Sub-Judge of Darjeeling.

Criminal Justice.

The judicial staff entertained for the administration of criminal justice consists of the District Judge of Purnea, the Deputy Commissioner, the stipendiary Magistrates stationed at Darjeeling, the Subdivisional Officer of Kurseong and the Deputy Magistrate of Silīgurī. Besides these, there is a strong bench of Honorary Magistrates at Darjeeling, some of whom are vested with first class powers and have authority to try cases singly, another at Kurseong, and a third in the Tarai, which is known as the Naksalbāri bench of Honorary Magistrates, but holds its Court at Bāghdogrā. One notable feature of these benches is that the Honorary Magistrates belonging to them are almost entirely Europeans.

The civil and criminal work of the district is, on the whole, very light. The hill people are seldom plaintiffs, as they are not litigious by nature; it would be impossible to find worse defendants, as they are so afraid of the terrors of the law that more often than not they will not defend a suit; and, as witnesses, they are practically worthless. They generally spoil even good cases by their inability to grasp what the aim of cross examination is; and the result is that, as a rule, they answer the astute pleader engaged on the other side in exactly the way he wishes them to, not apparently because they lie maliciously, but merely to please and soothe the troublesome pleader from the plains. As regards criminal cases, the Nepalese, who constitute a large portion of the inhabitants, are remarkably law-abiding, and both they and the other hill races are very different from the plainsmen in having an inherent dislike for litigation. Heinous crimes are comparatively rare, and the commonest offences are those against the person, such as affrays, assaults and cases of wounding, which are common among a people of considerable courage, a keen sense of honour, and a quick temper, with whom the *kukri* is the national weapon.

For the protection of life and property and the detection of crime the district is divided into 4 police circles or thānas, viz., Darjeeling, Jorbangalā, Kurseong and Siligurī, and, besides these, there are 4 independent outposts and 23 dependent outposts, including 14 patrol posts. There are therefore altogether 31 centres for the investigation of crime. The regular police force consisted in 1905 of a District Superintendent of Police, 5 Inspectors, 24 Sub-Inspectors, 2 Sergeants, 49 Head Constables and 351 constables, making in all 432 men. In addition to the regular police, there is a rural force or village-watch consisting of 24 *dafadārs* and 153 *haukīdārs*. Owing to the mountainous nature of the country and the difficulties of communication, the force is proportionately stronger than in other parts of the Province, there being one policeman for every 2·7 square miles and for every 576 persons. The cost of maintenance in 1905 was $1\frac{1}{4}$ lakh. POLICE.

In the hills the police are almost all Nepālis, recruited from among Jimdārs, Murmis, Mangars and Gurungs, who form a stalwart and alert body of men.

The District Jail is situated at Darjeeling, and there are small subsidiary jails at Kurseong and Silīgurī. The latter have accommodation for 24 and 8 prisoners, respectively, and are merely lock-ups in which prisoners sentenced to imprisonment for a fortnight or less are confined. The District Jail is an old-fashioned building, which can contain 130 prisoners; there are cells for 11 prisoners, and barracks without separate sleeping accommodation for 119 prisoners, including the hospital which contains 10 beds, a building reserved for 9 under-trial prisoners, and another intended for 4 European prisoners. The bakery, from which bread is supplied to the troops and the general public, constitutes the chief industry; but oil-pressing, bamboo and cane-work, carpentry and boot-making are also carried on. The average jail population is generally far below that for which accommodation is provided. JAILS.

CHAPTER XIV.

LOCAL SELF-GOVERNMENT.

LOCAL BODIES. THE Local Self-Government Act has not been introduced into the district, and there are consequently no District Board, Local Boards or Union Committees. Many of the duties discharged by ordinary District Boards devolve upon the Deputy Commissioner in his capacity as administrator of the Darjeeling Improvement Fund, a fund peculiar to Darjeeling, which is devoted to the general improvement of the district. As administrator of this fund, the Deputy Commissioner has extensive powers of control over local affairs, for it rests with him to see that proper provision is made for medical relief, for village sanitation and conservancy, for veterinary assistance, etc. The supervision of education rests to a certain extent with the District Committee of Public Instruction, a consultative body which is intended to assist the Deputy Commissioner in the general supervision of the Zilā school and the control of primary education, including the administration of the primary grant and the examinations for and the award of primary scholarships. The maintenance of the more important roads and bridges rests with the Public Works Department, but the remainder are under the control of the District Road Cess Committee. This body consists of 17 members, 5 official and 12 non-official, with the Deputy Commissioner as *ex-officio* Chairman. Under the District Committee there are two Branch Road Committees, the Kurseong Branch Committee composed of 9 members, of whom 3 are official and 6 are non-official, and the Tarai Branch Committee, with 6 official and 12 non-official members. The District Committee administers the amounts realized from the road cess, which amounted to Rs. 22,500 in 1905-06, and also receives a grant of Rs. 7,000 from the Darjeeling Improvement Fund as a contribution for the maintenance of communications. It is in direct control of the roads in the hills to the west of the Tīsta in the head-quarters subdivision, those to the east of that river being kept up by the ryots of the Kālimpong subdivision; and it makes allotments to the Branch Committees for the roads and bridges under their charge in the hills of the Kurseong thāna and in the Tarai.

MUNICIPALITIES. **Darjeeling.** The only municipalities are those of Darjeeling and Kurseong. The limits of the Darjeeling Municipality, which was first constituted in 1850, were originally co-extensive with those of the tract ceded by the Rājā of Sikkim in 1835, which covered an area of 138 square miles, and extended from the hills below Pankhābāri to the borders of Sikkim on the north. It had to provide funds for the upkeep of a conservancy and police establishment in the station of Darjeeling, and for the maintenance of 120 miles of roadway in the interior. The total amount available for this extensive charge was only Rs. 20,000 per annum.

Area. The area of the Municipality as now constituted is 4·85 square miles. It begins at Jorbangalā to the south, and extends to a point on the road to Takvār below St. Joseph's College on the north. On the east it is bounded by the Calcutta road and a strip of land below it. The boundary line then runs past and below the Chaurāsta and the Bhotiā Bastī, until it joins the boundary below St. Joseph's College. On the west it is bounded by the cart road and a strip of land below it, the boundary line continuing past and below the bazar through the Happy Valley Tea Estate, until it reaches the boundary below St. Joseph's College mentioned above. The mean length of the town from south to north is approximately 5 miles, and its mean breadth is roughly about 2½ miles.

Municipal law. The administration of the Municipality is governed by the Bengal Municipal Act and by a special Act [Act I (B.C.) of 1900], which was introduced in order to prevent the recurrence of such disastrous land-slips as those of 1899. Up to that time the municipal law in force in Darjeeling was the ordinary Municipal Act, III (B.C.) of 1884, which had been framed with reference to the requirements of towns in the plains, and was in many ways unsuitable to the circumstances of a town situated on a steep hill-side. It was found that the landslips were in many cases due to defects which the Municipality had hitherto no power to deal with, such as defective supervision of building sites and drainage, neglect to reduce or protect steep slopes, quarrying in unsafe localities, etc. A Bill was accordingly introduced, which passed into law as Act I of 1900, to extend the power of the municipality and to give it authority to take measures necessary to ensure the safety of the town. The most important additions to the powers of the local authorities conferred by this Act were the following. All roads, private as well as public, and all bridges were brought under control, in respect of their construction, maintenance and closure, and power was taken to enforce any alterations in them necessary to secure the stability of any hill-side or bank, or any buildings situated on them. Authority was given to enforce the repair and, in extreme cases, the removal of any building which threatened the security of a hill-side or bank, and to compel the owners to protect the sites when insecure. Similar provisions were enacted in respect of all drains, private as well as public; and power was taken to enforce the execution of schemes of drainage for groups of buildings or holdings in combination, the construction of revetments and retaining walls, the turfing of banks and the sloping of the hillsides to the angle of safety, whenever any of these works might be necessary for the general safety. A complete set of building regulations was also provided, which included full powers to regulate the excavation and preparation of building sites, and to prohibit building on any site considered insecure by professional authority. This was one of the chief dangers to be guarded against, and was the direct cause of much of the loss of life and property in 1899.

Administration. The affairs of the Municipality are administered by a Municipal Board consisting of 25 members, with the Deputy Commissioner as Chairman, of whom 3 are *ex-officio* members and 22 are nominated by Government. Nearly all the Municipal Commissioners are Europeans. For the purposes of administration the town is divided into 9 wards, and a Committee, composed of Commissioners, is appointed to report on and attend to the affairs of each ward. These Committees are not vested with specific powers, but all questions of importance arising within the respective wards are referred to them for enquiry and report, and their recommendations are then laid before the Commissioners in meeting.

Besides the Ward Committees, there are a number of consultative Committees, formed for the various departments of the administration, such as the Assessment, Appeal, Legal, Audit and Finance, Executive and Works, Sanitary and Conservancy, Hospital and Dispensary, Water-supply and Electric Light Committees. These Committees deal in the first instance with questions concerned with the various departments, and formulate proposals for the approval of the general body of Commissioners. The total number of rate-payers is 2,035, or 12 per cent. of the population in municipal limits.

Income. From the Statistical Appendix it will be observed that the average income of the Municipality for the decade ending in 1901-02 was Rs. 2,19,000. In 1905-06 the receipts, excluding loans, grants from Government and advances, amounted to Rs. 2,02,850, besides an opening balance of Rs. 19,660; including all these items, they aggregated Rs. 3,22,880. The main source of income is a rate levied on holdings in the town at 7 per cent. of their annual valuation; this is generally assessed on the rentals, but the holdings in or near the bazar and on the outskirts of the town, which are owned and occupied by natives, are usually assessed at a certain scale of rates fixed for every 100 square feet of ground area occupied. This rate brought in Rs. 55,250, and the next important source of revenue consisted of the rents of lands and houses owned by the Municipality, which amounted to Rs. 45,300. The Municipality owns 45 acres of land in the bazar, various buildings leased out as shops and residences, a Town Hall rented by the Amusement Club, and two covered markets, where sites are let out for the sale of meat, poultry, fish, vegetables, etc. In the same year the lighting rate, which is assessed at $2\frac{1}{2}$ per cent. on the valuation of the houses, yielded Rs. 38,450; the water-rate, assessed at $2\frac{3}{4}$ per cent., Rs. 22,590; and conservancy, including scavenging and latrine rates, Rs. 22,550; the latter rates are assessed at Re. 1-12 to Rs. 17 on houses valued up to Rs. 600 and at Rs. 3 per cent. on houses valued above Rs. 600. Besides these receipts, the municipal market brings in a revenue of over Rs. 9,000. Apart from loans and extraordinary items of receipts, the ordinary income of the Municipality is Rs. 2,18,390, and the current demand on account of taxes is Rs. 1,41,000. The incidence of taxation is higher than in any other town in Bengal, being Rs. 8-5-10 per head of the population.

The average annual expenditure in the decade ending in 1901-02 was Rs. 1,72,000, and in 1905-06 the total expenditure amounted to Rs. 3,01,800. Excluding debts and extraordinary heads, however, it was only Rs. 2,50,250; and of this amount, 26·9 per cent. was spent on lighting, 24·8 per cent. on conservancy, 17·4 per cent. on public works, and 7·6 per cent. on the water-works. The heaviest charge on the income of the Municipality is the electric-light installation, introduced in 1897, which supplies the streets and a large number of the houses. The installation is at Sidrapong, on the lower slope of the hill some 4 miles to the west of the town. The conservancy arrangements are on the whole very complete, a large staff of sweepers from Alwar in Rājputāna being entertained and the night-soil removed by train to Batasiā near Ghum. It is now proposed, however, to substitute for the present arrangement a complete system of septic tanks and sewage pipes. A volunteer fire-brigade has recently (1906) been started. Expenditure.

The length of the roads kept up by the Municipality is 37 miles; and the maintenance of these roads and of the various buildings is a heavy recurring charge. The water-supply is obtained from a number of springs on the western flank of Senchal, where it is led into settling tanks. It is then conveyed by a 9-inch pipe as far as Ghum, and thence by two lines to reservoirs built at St. Paul's School and Rockville. From these reservoirs it is distributed to the town through pipes of various dimensions having a total length of 12½ miles. It is in contemplation to introduce a filtration scheme in order to avoid all possible risk of contamination to the water. The liabilities of the Municipality are heavy, owing to the large amount of loans which it has from time to time received from Government for large schemes, such as the water-works and the electric-light installation. These loans amount to 6 lakhs of rupees, and nearly Rs. 35,000 have to be set aside every year for the payment of the principal and interest.

Kurseong. Kurseong was constituted a Municipality in 1879, and its affairs are administered by a Board consisting of 12 Municipal Commissioners, of whom one, the Civil Medical Officer, is an *ex-officio* member, while 6 are elected and 5 are nominated by Government. The area within municipal limits is 2 square miles, and the total number of rate-payers is 706 or 15·7 per cent. of the population. During the 10 years ending in 1901-02 the average annual income was Rs. 10,000; and in 1905-06 the income amounted to Rs. 72,200, but of this sum Rs. 51,600 represented a loan granted by Government. The principal source of income is a tax on houses and lands, which in that year brought in Rs. 10,250; and after this, the water-rate, which realized Rs. 4,150, and conservancy, including scavenging and latrine rates, the receipts from which amounted to Rs. 3,480. The other receipts are insignificant. The tax on houses and lands is a rate assessed on holdings at 7½ per cent. of their annual valuation; the water-rate is levied at 5 per cent. on the valuation of houses which have connection pipes, and at 2½ per cent. in the case of houses which have no connection pipes, but use water supplied by the Municipality; while the latrine fee is assessed at the rate of Re. 1-8 to Rs. 11 on houses valued up to Rs. 600 and at 2 per cent. of the valuation on houses valued above that amount. The current demand of taxes is Rs. 18,700, and the total incidence of taxation is heavy, being Rs. 4 per head of the population.

The expenditure during the decade ending in 1901-02 was on the average Rs. 9,600 a year, and in 1905-06 it amounted to Rs. 73,400, but of this amount Rs. 51,100 represented advances and sums invested in securities. Conservancy constituted the heaviest charge on the Municipality, accounting for 35·4 per cent. of the expenditure, and then came medical relief (13·3 per cent.), water-supply (13 per cent.), general establishment (11·55 per cent.) and drainage (10·2 per cent.)

The two principal requirements of the town at present are an effective system of drainage and an extended water-supply. The drainage has been rendered as complete as it could be with the limited funds at the disposal of the Municipality, but much yet remains to be done in putting the numerous *jhorās* in a more sanitary state and in having a regular system of drains to carry off the waste water and filth from the bazar to a suitable distance down the hill. The water-supply is obtained from 15 springs on Dow Hill above the town; it is collected in a reservoir with three tanks and is distributed from a reservoir by pipes having a length of 4 miles. There is, at present, a certain scarcity of water during the hot weather months; and it has accordingly been proposed to tap more springs in the forest above Sepoydhura, and thus to make the supply adequate for the increasing population and the numerous buildings being erected in the town.

CHAPTER XV.

EDUCATION.

NATIVE EDUCATION. No article on the progress of education in the district of Darjeeling would be complete or accurate which did not accord the first place of honour to those Christian missionaries who have been pioneers in this as in much other good work. Not only have the missionaries been the pioneers of education among the native population, but for the last generation practically the only organization for meeting the primary educational wants of the people of the hills has been the Church of Scotland Mission. Government has had its own Anglo-Hindī and High schools and has established and maintained several vernacular schools in the Tarai, but it has had such confidence in the Mission—a confidence which public opinion, as voiced by municipalities following the lead of Government by planters and by Indians of different classes and creeds, has emphasised as not misplaced—that up till this year (1906) all the Government contributions to primary education, as well as those of the two municipalities of the district, have been entrusted to the Mission for expenditure; for it has been found that practical solid educational work on broad wise lines has been obtained through the Mission at a minimum cost to Government.

Early efforts. When the British took over the district, popular education was practically unknown. A few of the better classes had, spasmodically, private tutors for their boys; a few, who themselves could read, tried to hand on their gifts to their families; and in Buddhist monasteries, novitiate monks were taught to chant Tibetan texts. But of education in general there was none and no schools worthy of the name were in existence. The first attempt to reach the hill people by education was made more than half a century ago by the Revd. W. Start, a private missionary, who added to his record of good work in Darjeeling by opening a school for Lepchās. After him came a band of German missionaries, one of whom, Mr. Niebel, devoted himself

I am indebted to the Revd. R. Kilgour, B.D., for this account of education among the hill people.

especially to school work, prepared some Lepchā primers, and gathered some boys together into schools.

But it was not till the advent of the Revd. William Macfarlane, in 1869, that any broad scheme of vernacular education was devised for the district. Mr. Macfarlane soon saw that the one thing required for the development of the district, for raising the people in the scale of civilization, and incidentally for obtaining a powerful lever for his mission work, was some statesmanlike scheme of education. A highlander himself, he set himself to devise for these Indian highlanders a system of education based on that of which he himself was an admirable product, the system of his native land, which aimed at putting some opportunity for even the initial stages of learning within the reach of every child. He realized that the first thing necessary to secure this end was a class of trained teachers; and with this object he collected a band of hill lads, to teach whom he devoted the first years of his missionary life in the hills. This class was the nucleus of the Training School at Kālimpong, which now trains teachers for the whole district. As he found that to take boys away for such a course of training meant a practical diminution in the income of the family, Mr. Macfarlane induced Government to offer scholarships for the students during their course. He grasped the fact that the Nepālis were people of a stronger character than any of the other hill tribes, and he soon found that their language was so akin to Hindī that he could use many Hindī text-books as a means of instruction. He also found that the Lepchās and Bhotiās, from their contact with Hindī and Nepāli-speaking peoples, were soon able to converse in this language. So he fixed upon it as the *lingua franca*, and in it prepared text-books, some of which are still in use in the district. He himself taught hour after hour in the face of many discouragements, great difficulties, and the frequent disappearance of the most promising pupils. But at last he was able to offer, with the help of Government, to start primary schools all over the district. Government at once came to his aid with liberal grants. He enlisted the sympathy of the tea-planters, and so won his way to the hearts of the cultivators and village headmen that soon many of them offered to build school-rooms, if he would only supply the teachers. Some of the tea gardens and some of the native cultivators also regularly contributed to the support of the teachers—a support continued in many of the old places and extended to several new schools up to the present time. **Mr. Macfarlane's work.**

In a few years Mr. Macfarlane's system had so taken hold of the district that in 1873 there were 25 Primary schools with 615 **Subsequent progress.**

boys and girls receiving instruction. The work, of which the foundations were thus laid broad and deep, has steadily grown under his successors, and the Educational Department of the Church of Scotland Mission work in the district is now a most important factor in the spread of knowledge amongst the people. There is a special training school for teachers, with a European Principal, in which 40 students are at present under training. Six missionaries spend part of their time in superintending the schools of the tracts allotted to their charge, arranging for new openings, appointing and looking after teachers, and regularly examining their work. Each charge has been mapped out into parishes under the immediate supervision of the pastor or catechist, who regularly inspects the schools in his parish, thus introducing the old Scottish parochial school system. There are now 70 schools, of which 55 are both day and night schools, with a roll of 2,420 boys and 300 girls, and an average attendance of 1,880. A recent return showed that very few centres of population are now without some school situated within a reasonable distance, in many cases within a mile or two. The schools are scattered throughout the whole district, in the towns, in the villages, and on the tea gardens; and the help rendered by Government, villagers and tea-planters, has gone on steadily since Mr. Macfarlane's day. The main educational developments since his time have been the addition of a few Anglo-Hindi schools, some Upper Primary schools, a Tibetan school and special girls' schools where such have been found necessary. But in the main the educational work is still conducted on the lines laid down by the pioneer.

Primary education. Government has its own Anglo-Vernacular schools at Darjeeling and Kurseong reaching a special class, principally the children of Bengali and other immigrants into the district; but the main educational work must always remain that of Primary schools, the object of which is to teach the hill children to read and write in their own vernaculars, though there is no objection to the chosen few passing beyond that stage. The tea-planters, who are the principal employers of labour in the district, would resent from an economic point of view, both for the sake of the employés as well as for their own, any attempt to make, for example, Anglo-Vernacular education general; and the missionaries are at one with them in this opinion.

The instruction given in the Primary schools is of the simplest, viz., reading, writing, spelling, arithmetic, a little geography, and a very little science. Few of the children proceed beyond the initial stages. Most are taken away to work, and thus add to the family income, just when they would be passing into the second or third book; so that a far greater number of scholars pass through the schools in a year than is shown in the figures at any particular date. In simple reading, etc., a hill child learns as quickly as most, but they are deficient in arithmetic.

Girls' schools have developed greatly during the past few years, principally owing to the advent from Scotland of ladies with a special aptitude for teaching. In Darjeeling there is a Girls' Boarding School and Female Teachers' Training School, teaching up to the Upper Primary standard and able to train for the teachers' certificates, with industrial departments producing much useful work. At Kālimpong there is a special Lace School under the charge of a lace expert paid by Government. Education of women.

Night schools have always formed a special feature of the Mission's educational work. These offer facilities for a class who, while out at work all day, desire to spend a few hours improving their minds at night. They are especially popular on tea gardens. Night schools.

As already stated, Government has up to this year expended all its grants for primary education in the hills through the agency of the Mission. But the time has come when further advances must be made, now that the ground has been prepared, and the initial difficulties have been overcome. The Mission authorities feel they have reached the maximum of the financial contribution they can make to this work; and Government, while recognizing the pioneer work it has done, proposes that a new system of Government schools should be started on those tea gardens which desire it. Government offers a grant-in-aid, and expects the garden to provide and maintain the school-house and bear a part of the cost. But it is still left open to the individual planter to ask for the help of the Mission in the superintendence of the schools—a help which, there is reason to believe, will not be refused; and this development should mark a great increase in the educational facilities of the whole district. Future possibilities.

The preceding account, which has been contributed by the Revd. R. Kilgour, B.D., shews how great the progress of education among the native population of the hills has been during the last half century; and it is only necessary to quote a few statistics for the district as a whole to illustrate its practical results. In 1860-61 there was only one school receiving Government aid, the English school at Darjeeling, which had a total attendance of 16 pupils; in 1870-71 one more English school and 16 vernacular schools had been started, and there were, in all, 18 Government and aided schools attended by 500 pupils; in Statistics of education.

1872-73 the district contained 3 Middle schools, 23 Primary schools, 1 normal and 1 girls' school, the number of pupils on the rolls being 680. In 1894-95 the number of schools had increased to 109, with an attendance of 3,830 pupils; and in 1904-05 there were altogether 142 schools with 3,950 pupils receiving instruction. Of these schools, two are High schools, of which one at Darjeeling had an attendance of 229 and the other at Kālimpong contained 70 boys; four were Middle English schools, and one was a Middle Vernacular school, the total attendance at these schools being 330; two were Training schools; five were Upper Primary schools, and 124 were Lower Primary schools with an attendance of 240 and 3,080 respectively. Altogether 19·8 per cent. of the boys of school-going age were receiving instruction, while the percentage of girls attending schools was far above the Provincial average. The total expenditure on education in the same year amounted to Rs. 38,216 only.

These figures do not at first sight appear to shew any very general diffusion of education among the people at large, judged by an European standard, but, as compared with the rest of the Province, Darjeeling occupies a high place. At the census of 1901 a special return was made of those who were able to read and write any language. People of whatever age who could do this were entered as literate, and those who could not as illiterate. The qualification seems a simple one, but, even so, only 1 male in 10 and 1 female in 200 was able to satisfy it in the whole of Bengal. In Darjeeling, however, it was found that no less than 12 per cent. of the males, or approximately 1 in 8, were literate, while the proportion of literate females rose during the decade ending in 1901 from 5 to 14 per 1,000 (approximately 1 in 71)—a ratio surpassed by no other district in Bengal or Eastern Bengal outside Calcutta. The advance of education among the female population is all the more satisfactory, when it is remembered that 30 years ago there was only one girls' school in the district, the existence of which was chiefly owing to Miss Macfarlane's energy and perseverance. The Nepālis, it was said, thought the idea of educating girls quite absurd, but some encouraging symptoms had appeared, as a number of lads who were trained in the normal school had begun teaching their sisters since they became teachers.*

Contrast between the hills and the Tarai.

The two most prominent features of the progress of education in the district are the very great advance which has been made in the hills in spite of the difficulties of teaching a polyglot population

* Statistical Account of Bengal, vol. X (1876).

scattered among the mountains and the backwardness of the Tarai. Writing in 1873 Mr. Macfarlane remarked :—" The boys in the Tarai schools are far ahead of those in the hills, as regards the progress made in school. It fact, as regards education, the Tarai seems to be very like the rest of the plains of Bengal. In the hills, everything—races, language, and condition of the people—is different. It is hard to say which of the two classes forming the mass of the hill population - the village agriculturists and the tea plantation coolies—presents the greater difficulty as regards education. Among the villagers, little Nepāli boys, almost as soon as they can distinguish between a goat and a sheep, are employed to look after their parents' flocks; and the teachers find it, as a rule, exceedingly hard work to collect half a dozen of them and keep them regularly at school. Again, on the tea plantations, from the end of March till the beginning of November, a little boy with scarcely strength enough to carry two or three seers on his back gets Rs. 3 a month in wages, so that during that period the plantation schools are almost entirely deserted. All that the hill people care for their boys to learn are the merest elements of reading, writing and arithmetic. It is so easy to learn to read and write Hindī, that a sharp boy acquires a fair knowledge of both by the time he has gone through the Hindī primer. Accordingly we find that many boys, as soon as they have gone through the primer, consider their education finished. They can read and write to their own and their parents' satisfaction, and that is all they care for."

These remarks sufficiently shew the difficulty of bringing home to the hill people the benefits of even a rudimentary education, and they also show that 30 years ago the inhabitants of the hills were far behind those of the Tarai. Conditions have now been reversed. In the Tarai there are only 40 primary schools with an attendance of 660 boys, the percentage to those of school-going age being approximately 11 per cent. In the rest of the district there are 93 primary schools with 2,660 boys on the rolls, and approximately 19 per cent. of the boys of school-going age receive instruction. The backwardness of the Tarai is shewn even more clearly by the statistics obtained at the census of 1901, when it was found that in the Tarai only 1 person in 35 could read and write, and that in the rest of the district 1 person in 12 could do so. It may perhaps be contended that the latter figures are vitiated by the fact that they include the European population of Darjeeling and Kurseong, but if we take those for Kālimpong only, where the population is entirely native, with the exception of about a score of Europeans, the contrast is almost as great, for even there 1 person in 17 can read and write. In the Tarai, it may be added, primary education is managed directly by Government, and the Church of Scotland Mission is not its agent, as it is in the hills. Till 1874 all the primary schools in the Tarai were managed by it, but in that year the Mission, manned by only one European missionary and a very few native agents, gave them up; and when in 1895 the Mission proposed to resume the management of primary education in the Tarai, the proposal was not accepted.

Organization. With the exception of the Darjeeling High School, the schools in the hills are under the control of the Church of Scotland Mission, which receives fixed grants for their management from Government; one Sub-Inspector of Schools, who is under the Deputy Commissioner, has recently been appointed for the purpose of inspecting the primary schools. In the Tarai the education of the people is directly under the control of the Education Department, but the Deputy Commissioner, assisted by a Sub-Inspector of Schools, is the administrator of the primary grants and has power to appoint, punish, transfer or remove the *gurus*. The duty of the Sub-Inspectors is to inspect the primary schools; these officers belong to the Education Department, the nomination and transfer of the officers appointed to the post resting with the Director of Public Instruction, but they are under the general supervision of the Deputy Commissioner.

District Committee of Public Instruction. In the administration of education the Deputy Commissioner is assisted by the District Committee of Public Instruction. This is a body which has been constituted for the general supervision of education in districts where the Bengal Local Self-Government Act is not in force, under an order which lays down that in such districts the Magistrate, associating with himself the District Committee as a consultative body, should retain general control over all that concerns primary education, including the administration of the primary grant and the examination for, and award of, primary scholarships. The Commissioner is the President and the Deputy Commissioner is the Vice-President; as such he is the active head of the Committee. Beside assisting the Deputy Commissioner in matters of primary education, it is the function of the District Committee to see to the general supervision of the Zila school, so far as regards the maintenance of the buildings, the financial affairs of the school, and the settlement of questions of discipline that may be referred to them by the Head Master of the school; in all matters, in fact, except the determination of the course of instruction, and the appointment or promotion of teachers.

A large portion of the work of this Committee consists of the administration of the Darjeeling Zila or High School. The nomination of masters and regulation of studies rest with the Director of Public Instruction, but the discipline and management of the interior economy of the school rests with the Committee and the Deputy Commissioner. This school used to be known as the Bhotiā School and was formerly a Middle English school, but was raised to the status of a High school in 1892. It is open to all natives, and a great portion of the students consist of the children of immigrants from the plains, such as the native Government clerks and other Bengali and Hindustāni residents in Darjeeling. The remainder are chiefly Nepālis, but a small number of the pupils are Bhotiās and Lepchās, who do not intend to read for University examinations, are educated free, and are trained as explorers, interpreters and surveyors. The school is practically divided into two departments, one being a High school, and the other an elementary school for boys belonging to the hill tribes. **Darjeeling High School.**

The European schools in the district are secondary schools, based on the European model, which are mainly used by the classes of the community next below the class which sends its children to be educated in England. These schools are St. Paul's School (Church of England) and St. Joseph's College (Roman Catholic) at Darjeeling, and the Victoria School at Kurseong, an undenominational school maintained by Government. All these are boys' schools. The schools for girls are the Diocesan Girls' School (Church of England), the Loreto Convent (Roman Catholic), the Queen's Hill School (American Methodist), all situated at Darjeeling, and the Dow Hill Girls' School, maintained by Government, at Kurseong. **EUROPEAN EDUCATION.**

The history* of St. Paul's School is a very interesting one, and affords a striking instance of the vicissitudes and difficulties which have been the lot of many of the European schools in India. In the year 1830 Archdeacon Corrie and his friends established a High School for Europeans in Calcutta. The institution failed, and in 1846 it was broken up and replaced by the St. Paul's School, which was located in 1848 in a building in Chowringhee Road. For a time this school flourished, but it was unendowed, and in the end it failed to hold its own against its endowed rivals, the Doveton and La Martinière Colleges. It fell into debt, and in 1863 the school was closed and the premises **St. Paul's School.**

* See Progress of Education in India, 1897-98 to 1901-02, by R. Nathan, C.I.E., I.C.S.

were sold for a sum of Rs. 1,30,000. It was decided by the trustees to use the sale-proceeds to establish a school at Darjeeling, which would, it was thought, be a great boon to many of the inhabitants of Calcutta and Bengal. This was the time when Bishop Cotton was advocating the establishment of hill schools for Europeans. He raised funds by private subscription, and Government made an equivalent contribution. By this means a sum of Rs. 1,12,300 was collected, and it was invested in 1868 in the form of an endowment. The school was opened at Darjeeling in 1864 with 30 pupils on the rolls, and from that date until about the year 1895 the numbers gradually increased. The prosperity of the school then declined; it suffered from a growing tendency on the part of Anglo-Indian parents to send their children home to be educated, and also from the competition of the Roman Catholic St. Joseph's College. The number of pupils now averages a little over 100.

The school is managed by a Committee of which the Bishop of Calcutta is the President; and the staff consists of six masters, all of whom are graduates of Oxford or Cambridge. The course of study is that prescribed by the Cambridge University Local Examination Syndicate with the addition of religious training upon the principles of the Church of England. Preparation is also given for the various competitive examinations in India, *e. g.*, Rurki, Opium, Forest, Survey, Accounts, and also for the examinations of the English Universities, the Indian Civil Service, Woolwich, Sandhurst, etc. The school is frequented by the sons of Government officers and clerks, planters, railway officials, and the like. The ages of the pupils vary from about 8 to 19 years; as a rule, they complete the primary course by about 13, the middle course by about 15, and the Senior Cambridge Local course by about 18. About 40 per cent. of the boys go to England from the school—some when they reach the age of about 11 or 12, and others at or near the end of the course. The best pupils often enter Government service, *e.g.*, the Police, Public Works, Accounts, Survey and Opium Departments. Briefly, it may be said that the aim of the school is to provide, at as reasonable a charge as possible, education of a high standard on the lines of that given in an English public school, to the sons of members of the European community who cannot afford to send their children to England.

St. Joseph's College.

St. Joseph's College is a large Jesuit institution, the teaching staff of which is composed mainly of members of the Jesuit order who are unsalaried. It was opened in 1888, under the direction of the Fathers of the Society of Jesus, in a building known as St. Joseph's Seminary, Sunny Bank; and was transferred in 1892 to its present site at North Point. When first opened, there were only 35 boys on the rolls, but the number increased to over 200 in the year 1895, and that is still its average strength. The College aims at a sound knowledge of English, Mathematics, History, Geography, Modern Languages and the Classics. The study of English occupies the first place; Latin and French come next; then Mathematics, History and Geography. It is divided into 4 departments—the Primary, Middle, Higher and Special Departments. The Special Department has been opened to prepare young men for the different Government examinations, such as Accounts, Police, Opium, Forest, etc., as well as for entrance into the Engineering College at Rurki. Facilities are afforded for learning modern languages, music and drawing, and a complete apparatus for gymnastics is at the disposal of the pupils. A museum of physical science instruments and a laboratory for chemistry are attached to the College.

The Victoria School, Kurseong.

The Victoria School, Kurseong, is the only Government institution of its class in India. It was originally established in the year 1879 for railway employés, and its scope was afterwards extended to Government servants of all classes and to the children of officers paid out of local funds. The children of non-officials are also admitted, but they pay a higher rate of fees. At the end of the year 1905-06, it had 190 pupils on the roll, all of whom were boarders. It has a staff of 13 teachers, including a Gymnastic Instructor, six of the teachers having English or Irish qualifications, and two being graduates of the Calcutta University. The staff also includes a steward, two matrons, a trained nurse, and a lady house-keeper. The school has a Technical Department, affiliated to the Sibpur Civil Engineering College, in which a course extending over two years is provided for boys who have passed the Middle School examination. There is also a Commercial Department open to boys who have passed through the Middle School; the course extends over three years. It is proposed to erect a large preparatory school in the near future, close to the school and under the same management.

Girls' Schools.

The Diocesan Girls' School at Darjeeing has about 80 pupils, nearly all of whom are boarders. It has a staff of 12 mistresses and teachers, including teachers of music, besides 3 matrons. The governing body is the Diocesan Board of Education, consisting of the Bishop and Archdeacon of Calcutta, and the direct management of the school is entrusted to two sisters provided by the Clewer Sisterhood in England. The Loreto Convent has 165 pupils, the majority of whom are boarders. The staff consists of 22 nuns and 5 lay teachers. The Dow Hill School is the girls' department of the Government school at Kurseong. It has 120 pupils, all of whom are boarders, and a staff of 9 class mistresses and 2 music teachers. The Subordinate staff consists of 2 matrons, a nurse and a lady house-keeper. There is also a Training College attached to the school, with 20 students and a staff of 2 teachers.

CHAPTER XVI.

GAZETTEER.

British Bhutān.—A name formerly given to the portion of the district lying to the east of the Tīsta, which was wrested from Bhutān by the war of 1864-65.

British Sikkim.—A name formerly given to the hill territory of Darjeeling. The term was originally applied to the hill territory ceded by the Rājā of Sikkim in 1835, which formed the nucleus of the present district. This tract, which included an area of 138 square miles, consisted of an irregular strip of land, about 24 miles long and from 5 to 6 miles wide, extending from the foot of the hills below Pankhābāri on the south to the frontier of Sikkim on the north. The boundaries of British Sikkim were extended in 1850 by the annexation from Sikkim of three tracts in the hills, viz., the territory in the north-west beyond the Little Rangīt river, and two strips of land extending on either side of the old hill territory to the Tīsta on the east and to the Nepāl boundary on the west. At the same time, the British also annexed the territory at the foot of the hills, which was then known as the Sikkim Morung and is now called the Tarai.

Dāling.—An old Bhutanese fort situated in the south-east of the Kālimpong tract to the east of the Tīsta. The fort, which was stormed by the British troops in the Bhutanese war of 1864-65, is situated at a height of 3,350 feet above sea-level on a lofty eminence overlooking the Duārs. It has been dismantled, and only a few picturesque ruins remain.

Dālingkot.—A name formerly given to the portion of the district east of the Tīsta. The name, which is derived from that of the fort mentioned above, was given to this tract, when it was constituted a subdivision in 1865.

Darjeeling.—The head-quarters of the district situated in the Lower Himālayas in 27° 3′ N., and 88° 18′ E., at a distance of 379 miles by rail from Calcutta. The name Darjeeling is a corruption of Dorje-ling and means the place of the *dorje*, the mystic thunderbolt of the Lāmaist religion, a designation formerly given to the Buddhist monastery which used to stand on the top of Observatory Hill.

The town is situated on a long spur, which projects to the north from the Senchal-Singālilā range of mountains. This spur rises somewhat abruptly from Ghum to an elevation of 7,886 feet at Katāpahār, and then gradually descends to 7,520 feet at Jalāpahār and to 7,002 feet at the Chaurāsta, which may be regarded as the centre of the station. It rises again to 7,163 feet at Observatory Hill just above the Chaurāsta, and then divides into two, the Lebong spur and the Takvār spur, which sink down to the valley of the Rangīt. In shape, Darjeeling resembles the letter Y, the base being represented by Ghum, the capital or upright portion of the letter by the ridge stretching from Katāpahār to the Chaurāsta, and the two arms to right and left by the spurs on which stand Lebong (5,970 feet) and Birch Hill, the highest point of which is 6,874 feet above sea-level. The total area of the station is nearly 5 square miles, and the difference in height between its highest and lowest points is about 2,000 feet, Katāpahār being 7,886 feet and Lebong 5,970 feet high. The ridge is very narrow at the top, along which some of the European houses are perched, while others occupy positions on its flanks. The eastern slope which looks down into the Rangnu valley is very steep, but the western slope is much gentler, and it is here that most of the public buildings and the bazar have been built. Below the latter lie the jail, the Botanic Gardens and the native town, which is chiefly occupied by the poorer classes of natives, and consists of huts built without method or regularity. Still lower down are tea gardens, which come up to the limit of the houses.

The snowy range. The situation of Darjeeling is singularly beautiful, standing as it does on a narrow ridge that juts out into a vast basin in the heart of the Himālayas. To the west and south it is enclosed by mountains higher than itself, but to the north and north-east the view is more open, and the eye travels past range upon range of hills, rising in ascending waves behind the cultivated slopes of the nearer Sikkim hills, until the prospect terminates in the distant snowy mountains with their long girdle of rugged peaks. The spectator, in fact, stands on the stage of a vast amphitheatre of mountains, the sides of the amphitheatre being formed by the Singalīlā chain 20 miles to the west and by the loftier Cholā range 40 miles away on the east. To the extreme right are the mountains of Bhutān, conspicuous among which are Gipmochi (14,518 feet) and the fortress-like crest of Chumanago or Dopendikang (17,310 feet); to the left are the mountains of Nepāl and the long range marking the limit of British territory, on which stands Sandakphu, (11,929 feet); in the middle are the lofty mountains of northern Sikkim. In front, at a distance of only 45 miles, the great twin peaks of Kinchinjunga (28,146 feet), tower above the titanic group of snowy mountains which fills the northern horizon. This gigantic mountain is flanked on the west by Kabru (24,015 feet) with its crest resembling that of a double-poled tent, and by the sharp horn-like peak of Jāno (25,300 feet); and on the east by Pandim (22,020 feet), a name meaning the king's minister, which has been given to this peak because it stands by the side of the king of mountains, Kinchinjunga. Further to the east is a long line of other snowy mountains, of which the sharp conical peak of Narsingh (18,145 feet) and the graceful snow-mantled crest of Siniolchu or D2 (22,520 feet) are the most beautiful; Kinchinjhau (22,720 feet), with its flat table-like top showing at a distance of 69 miles, is the most curious; and the great mass of Dongkya 72 miles away is the most distant. In the near foreground the cone-shaped peak of Tendong (8,676 feet) stands out as the most prominent feature in the landscape. It is a sacred mountain to the Lepchās, by whom it is regarded as the Mount Ararat of their race. According to their tradition, there was a great flood in the days when there were none but Lepchās in the land. The few survivors fled to the high crest of this mountain, which miraculously elongated itself as the flood rose, and thus kept its clinging refugees from being engulfed. Its name Tendong or "the uplifted horn," still preserves the memory of this ancient legend. It is a curious coincidence that this mountain should have been called Mount Ararat by the early residents of Darjeeling, but this is only due to the fact that the peculiar shape of its summit resembled that of the peak shown in children's books as the resting-place of the Ark.

Many descriptions have been given of the beauties of the snowy range as seen from Darjeeling; but perhaps none is so pleasing as that which Sir Joseph Hooker gave while protesting that it was impossible for him to give any description at all. "The most eloquent descriptions I have read," he says, "fail to convey to my mind's eye the forms and colours of snowy mountains, or to my imagination the sensations and impressions that rivet my attention to these sublime phenomena when they are present in reality; and I shall not therefore obtrude any attempt of the kind upon my reader. The latter has probably seen the Swiss Alps, Scenery.

which, though barely possessing half the sublimity, extent, or height of the Himalaya, are yet far more beautiful. In either case he is struck with the precision and sharpness of their outlines, and still more with the wonderful play of colours on their snowy flanks, from the glowing hues reflected in orange, gold and ruby, from clouds illuminated by the sinking or rising sun, to the ghastly pallor that succeeds with twilight, when the red seems to give place to its complementary colour green. Such dissolving views elude all attempts at description, they are far too aerial to be chained to the memory, and fade from it so fast as to be gazed upon day after day, with undiminished admiration and pleasure, long after the mountains themselves have lost their sublimity and apparent height. On first viewing this glorious panorama, the impression produced on the imagination by their prodigious elevation is that the peaks tower in the air and pierce the clouds, and such are the terms generally used in descriptions of similar alpine scenery; but the observer, if he look again, will find that even the most stupendous occupy a very low position on the horizon, the top of Kinchin itself measuring only 4° 31′ above the level of the observer. Donkia, again, which is 23,176 feet above the sea, rises only 1° 55′ above the horizon; an angle which is quite inappreciable to the eye, when unaided by instruments."

Unfortunately this beautiful panorama is only too rarely unveiled, except in the hot-weather months, at the close of the rains, and in the beginning of the cold weather. At other times, it is more often than not hidden by envious mists, which lift only at rare intervals for a few minutes at a time. Throughout the greater part of the year, the prevailing wind is from the south-east and comes laden with moisture from the Bay of Bengal. It rises at sunrise, and its vapours are early condensed on the flanks of Senchal; billowy clouds rapidly succeed small patches of vapour, which, rolling over to the northern side of the mountain, are carried north-west over a broad intervening valley to Darjeeling. Very often the whole eastern sky, from the top of the Darjeeling ridge, is enveloped in a dense fog by 9 A.M., while the western exposure enjoys sunshine for an hour or two later. Before noon, however, masses of mist roll over the ridge to the westward, gradually filling up the valleys, so that the whole station is soon enveloped in blinding mist and fog.

The station. The station itself is picturesquely situated on a long narrow ridge, with great valleys, 4,000 to 6,000 feet deep, on either side and a number of small spurs projecting from their flanks; but the lover of the beautiful will find little to admire in the medley of houses with corrugated-iron roofs which climb the hill-sides. Much, however, has been done to relieve their bare outlines, and it is only a question of time before the trees which have been planted will to some extent replace the forest growth which once covered the slopes. The taste of the pioneers of arboriculture in Darjeeling has, however, not always been very happy; and the predominant tree is the quick-growing cryptomeria which has been introduced from China and Japan. The seeds were obtained from trees that had been raised from seed brought to India by Mr. Fortune on his first mission to China in search of tea seed. On his return, he gave a quantity of seeds to Dr. Falconer, Superintendent of the Botanical Gardens, for distribution in the different hill stations; and a number of trees were raised in a nursery at Jalāpahār by Dr. Anderson during 1864-65 from the seed sent to Darjeeling. They have a somewhat stiff, funereal appearance, and have a curious resemblance to the conventional wooden trees which find a place in children's boxes of toys. Their stiffness is, however, relieved by a number of other trees which are found all over the station, such as birches, alders, maples, oaks, rhododendrons, the glossy-leaved *pīpli* (Bucklandia populnea), the scarlet-blossomed Erythrina, and the feathery blue-stemmed Sikkim bamboo; while the banks and grassy slopes have been planted out with dahlias, lilies and hydrangeas, and, in places, with the homely furze and primrose. Unfortunately, the deodar and other Himalayan pines do not thrive at Darjeeling owing to the excessive humidity which prevails.

Population. In 1872, before the construction of the railway, when the only approach to Darjeeling was by a long tedious march over the plains of Bengal, the population numbered only 3,157; but during the next nine years it increased by more than 100 per cent. In 1881 the construction of the Darjeeling-Himalayan Railway brought the station within a day's journey of Calcutta, and the population again doubled itself between the years 1881 and 1891. Its growth since 1891 has been less rapid, the disastrous landslips of 1899 having caused a temporary check to its development; but when the census of 1901 was taken it was found that the number of inhabitants had increased by 20 per cent. and amounted to 16,924. An enumeration carried out in the preceding September showed, however, a population of 23,852 or 50 per cent. more than in the cold-weather months, the difference being due to the fact that, during the hot weather and rainy months, Darjeeling is the head-quarters of the Bengal Government, and is crowded with visitors who escape to its cool bracing climate from the stifling heat of the plains. Its position as a summer resort for visitors and invalids is now firmly established, but the further extension of the town is a question of some difficulty. Most of the available building sites have been already taken up and built upon; the picturesque wooded spur of Birch Hill, where sites could be found, has been reserved as a park; and there appears at present to be little prospect of further expansion unless a suburb is built out in the direction of Ghum.

Observatory Hill. The most picturesque point in Darjeeling is Observatory Hill, which commands a magnificent view of the mountains and valleys by which Darjeeling is surrounded. Formerly, the hill was crowned by a Buddhist monastery, which was destroyed by the Gurkhas when they overran the country in the early part of the 19th century. This monastery, which was a branch of the Dāling monastery in Sikkim, was rebuilt on its old site above the cave which is reputed to lead underground to Lhāsa; but it was subsequently removed to the Bhotiā Bastī lower down the hill to the spot where it still stands. The old site is however still a place of great sanctity, at which the Buddhists make quaint offerings and libations, often brought in enamelled iron kettles made in Birmingham, and set up poles and bamboos fluttering with strings of flags, "the horses of the wind," which transport their prayers to the gods. Round this hill runs the Mall, a narrow circular road about 1 mile long, where pony races and gymkhānās used to be held. The monastery or *gompa* in Bhotiā Bastī, some hundred feet below, is perhaps one of the most interesting sights in Darjeeling. It is a simple two-storied structure, surrounded by tall posts bearing long strings of flapping prayer flags, with a number of small prayer wheels on each side of the entrance and two gigantic barrels of the same kind in the portico. The interior consists of two dark low-roofed rooms, in which are enshrined the idols and sacred vessels, painted scrolls, and a collection of Buddhist scriptures. Above it by the side of the road is a massive white-washed *chorten* erected to the honour of some pious Lāma.

The bazar. On the other side of the hill is the bazar, ugly enough in itself with its closely packed huts and shops, but of interest on account of the medley of races found in it. Here may be seen sleek, black-bearded Mārwāris, sitting in their cloth shops and perpetually conning mysterious account-books, before a little wooden door concealing the shrine of their favourite god Ganesh; specious smooth-tongued Kashmīri and Punjābi merchants; petty Hindustāni shop-keepers from the plains; and a crowd of hill people of various nationalities, such as the brawny Sikkimese Bhotiās, the placid Lepchās, and the active and alert Nepalese. The shops are equally varied in character, ranging from the ordinary glass-fronted shops, dealing in European piece-goods, groceries, glass, hardware and crockery, to dingy stalls containing a curious assortment of Oriental articles intended for visitors, such as turquoise, coral and amber ornaments, jade and agate cups and beads, Nepāli *kukris*, Bhotiā and Lepchā knives, brass tea-pots, prayer-wheels, bells, amulets and other curiosities illustrative of Buddhist monastic life.

Botanic Gardens. Below the bazar are the Lloyd Botanic Gardens, so called because the site was granted by Mr. Lloyd, the proprietor of Lloyd's Bank in Darjeeling. It was transferred to its present site from Rangarun in 1878, and some of the original trees, and many of the cryptomerias with which the bare slopes were planted up still remain. The garden is divided into two main parts, the upper or indigenous section, and the lower or exotic section ; but some small sections are reserved for the flora of the hills of Southern, North-Western and Eastern India. From its first establishment, the garden has done very useful work in introducing into the district many new plants, trees and shrubs, mostly from the temperate parts of the world ; but the number of species tried and found to be unsuitable to the climate of Darjeeling is greatly in the majority. The grounds, which are laid out in slopes, lawns and flower-beds, have a park-like appearance. They contain an extraordinary variety of different species in a small area. Among European species which have been introduced are the hawthorn, blackberry, gorse, broom, horse-chestnut, ash, elder, myrtle, birch, elm, yew, box, privet, clematis, lilac and honey suckle. From China and Japan come cryptomerias, plums, cherries, magnolias and maples, and from other parts of Asia the weeping willow and deodar. Africa is represented by many Cape bulbous plants and the cypress known as Callitris Whytei; America by the Magnolia grandiflora and the Juniperus Bermudiana; while the silver and black wattle, the silver oak, the blue gum or eucalyptus have been imported from Australia and New Zealand. In the centre of the garden is a Natural History Museum, which has been recently started and which contains a small but good collection of the birds, moths and butterflies of Darjeeling. A short distance to the south are the Victoria waterfalls, where a small stream, called the Kāgjhorā, which drains a large ravine above, comes down in a cascade over a precipitous crag some 100 feet high, and then hurries down to the valley below.

Birch Hill. A great contrast to the trim lawns of the Botanic Gardens is afforded by the natural beauty of the woods situated on Birch Hill. These woods, with their wealth of vegetation, shady walks and old moss-covered and creeper-bound trees, show what Darjeeling was like when it was first discovered. This is the only part of the station where the fine forests which once covered the hill-side have not been ruthlessly swept away by house-builders, though a few survivors of the forest giants may still be seen here and there.

Just below Birch Hill Park is the Cemetery, laid out in terraces on the steep hill-side. It contains the grave of General Lloyd, the discoverer of Darjeeling, who died here in 1865 just 30 years after he discovered the place; and close by is a tombstone erected to the memory of Csoma de Koros, an Hungarian philologist, whose life-long aim it was to prove by philological researches that the nomad races from whom the Hungarians are descended came from Tibet. When a young student, he begged his way across Asia and spent many years in a Tibetan monastery, compiling a great dictionary and grammar of the Tibetan language. On his way to Lhāsa to resume his labours, he died at Darjeeling in 1842. The saddest corner in the cemetery is that containing the tombstones erected in memory of the 10 children who were killed by the landslips of 1899, two dying in the arms of their faithful Indian nurse, while another inscription records in simple language how four out of six brothers and sisters who were killed by the slips were buried by God himself under the mountains on the other side of the hill.

Buildings. The buildings in the town are all modern and of little interest. The principal are the Shrubbery, the summer residence of the Lieutenant-Governor of Bengal, which was built on the site of an old cottage in 1879, and the Darbār Hall attached to it, which was added later; the Eden Sanitarium built in 1882 at the instance of Sir Ashley Eden, then Lieutenant-Governor of Bengal, on a site formed by cutting down the old Post-office hill on which the Bhotiā school stood; and the Lowis Jubilee Sanitarium erected in 1887 on land given by the Mahārājā of Cooch Behār. Of the churches the most prominent is St. Andrew's Church belonging to the Church of England. The present structure is modern, being built in 1870, but it occupies the site of the first church in Darjeeling, the foundation stone of which was laid on St. Andrew's Day, 1843. The modern building contains some memorial tablets which are illustrative of the history of Darjeeling, such as that to the memory of the soldiers who died on service in Sikkim in 1888-89, and that of General Lloyd recording the fact that "to his exertions and personal influence with the Raja of Sikkim the Province of Bengal is indebted for the Sanitarium of Darjeeling." The other churches are the Union Chapel of the Nonconformists, which was established in 1869 as a place of worship where ministers of any section of the Protestant church would be welcome to conduct service according to the views held by them; the Scots' Kirk built in 1894 by the Church of Scotland Mission; and the Chapel of the Immaculate Conception, close to the Roman Catholic Presbytery and Loreto Convent. The principal civil public buildings are the Bengal Secretariat Offices built in 1898; the *Kaçhahri*, containing the offices of Deputy Commissioner, District Superintendent of Police and Conservator of Forests, the Registration office, treasury and various civil and criminal courts, which was erected in 1897 in place of the old *Kachahri* which had been burnt, with nearly all the records, in the preceding year; and the jail, several hundred feet below the bazar, from which bread made by the prisoners is supplied to the troops and general public.

The chief educational institutions are St. Paul's School, which was established in Calcutta in 1845 for the education of Europeans and Eurasians and was removed to Darjeeling in 1864, and the St. Joseph's College, under the management of the Jesuit Fathers, which was founded at Sunny Bank in 1888, but was transferred to the present buildings at North Point in 1892. They are situated at extreme ends of the station, St. Paul's School being close to Jalāpahār at a height of 7,376 feet, and St. Joseph's College about 2 miles north of the railway station on the Lebong road, at an elevation of 6,507 feet above sea-level. There are two girls' schools for Europeans and Eurasians, the Diocesan Girls' School for Protestants and the Loreto Convent School for Roman Catholics. The principal school for natives is the Darjeeling High School, which contains a large number of pupils of different races, Nepālis, Bĕngalis, Hindustānis, Bhotiās and Lepchās. Education.

The most important medical institution is the Eden Sanitarium, where there is accommodation for 70 sick and convalescent patients in the main building and for 18 persons in the contagious wards; while the hospital, which was opened in 1901, contains seven beds and an excellent operating theatre of the most modern type. The Lowis Jubilee Sanitarium which is intended for natives provides accommodation for 99 persons; a phthisical ward, which is to be attached to this institution, is now under construction. The town also contains a large dispensary, the Victoria Memorial Dispensary, containing 45 beds, which does much useful work among the poorer Natives and Europeans. Medical.

The cantonments are situated on the ridges of Katāpahār and Jalāpahār above the town and on the spur of Lebong below it. At Katāpahār are Artillery barracks, which accommodate a battery of Field Artillery and a company of Garrison Artillery during the summer. Jalāpahār is a convalescent Depôt and contains Cantonments accommodation for 400 men. The Lebong cantonment below Darjeeling is occupied by one battalion of a British Infantry Regiment. The head-quarters of the Northern Bengal Mounted Rifles are at Darjeeling; the force consists of six companies with head-quarters at Kurseong, Jalpaigurī, Dām-Dim, Nagrakot, Alīpur-Duārs and Purnea, 3 companies of cadets and one reserve company.

In concluding this sketch of Darjeeling, it may be of interest to mention its appearance in 1840. The only public building was the Superintendent's *kachahri* just below Observatory Hill, a plain wattle and dab bungalow with an iron roof. There was no church and no clock, and it was suggested that, if Government could not afford a clock, a good sun-dial would be very acceptable! Only about 30 private houses had been built, all of wattle and dab, some with iron and others with bamboo roofs, mere cottages of a better sort. A drive 16 miles long had been made round the station, and the Auckland road had been laid out, this being described as "a splendid road leading to two magnificent waterfalls called the Eden Falls, and opening out extensive and imposing views as it winds along its varied and picturesque scenery": this road and the falls were apparently named after the Governor-General, Lord Auckland, and his sisters. The names still survive, and so does that of General Lloyd's house, Mount Pleasant. Dr. Campbell, the first Superintendent and the real founder of the prosperity of the district, lived at a place called One Tree, which is now occupied by the Beechwood estate; and just above on the side of the hill was the house of Lord Napier of Magdala, then a young Engineer officer, who was engaged in laying out the station and in constructing a road from the plains. To these three Darjeeling is indebted for its development; to General Lloyd for discovering the place and securing its cession; to Dr. Campbell for the introduction of tea cultivation and the general development of the district, hitherto inhabited by only a few aboriginal races; and to Lord Napier of Magdala for removing its inaccessibility and laying out the station.

Darjeeling subdivision.—The head-quarters subdivision of the district, lying between 26° 52′ and 27° 13′ N., and 87° 59′ and 88° 53′ E. The total area of the subdivision is 726 square miles, and it is divided into two parts by the river Tīsta. To the east is the Kālimpong tract, where nearly all the land not already reserved for forests is set aside for Lepchā, Bhotiā and Nepāli cultivators. To the west of the Tīsta the ridges and slopes above 6,000 feet and below 3,000 feet are generally covered by forests; and practically the whole country between these levels is under tea, with the exception of a few Government estates, the area reserved for cinchona cultivation near Sarail and Mangpu, and the tract to the north-west of the Little Rangīt, where the land is cultivated with ordinary country crops. The population, which was 133,386 in 1901, is contained in one town, Darjeeling, and in 181 villages. The density of the whole subdivision is 184 persons to the square mile; but it is far less in Kālimpong, where the tea plantations account for only 10 square miles, than in the western portion of the subdivision, where there is a large tea-garden population.

Ghum.—A village situated to the south of Darjeeling at a distance of 4 miles by the Cart road, but, if the more direct road over the hill is followed, the distance is only 3 miles. The village, which is more generally known as Jorbangalā among the natives, is perched on a narrow saddle which connects the Senchal range with the Darjeeling spur; and the consequence is that, being in a gap between the hills, it is an extremely rainy, misty and wind-swept place. It is a trade centre of some local importance, as it stands at the junction of three main roads, the Cart road from the plains, the road to the Tīsta Valley, which is the main route from the valleys, and the road leading to the Nepāl frontier. The village contains a Buddhist monastery, a police-station and railway station. The latter, which is some little distance above the bazar, stands at a height of 7,407 feet above sea-level, and is the highest point reached by the railway. Four miles to the west along the road leading to Sukiāpokhri there is one of the natural curiosities of the district, an enormous detached rock of gneiss standing 100 feet high, from which a magnificent view can be obtained of the Bālasan valley. According to local legend, criminals used to be executed by being hurled over this rock in the days when the hill tribes still ruled the country.

Hope Town.—*See* Sonāda.

Jalāpahār.—A hill just above the station of Darjeeling which forms part of the spur running from Ghum down to Lebong. Jalāpahār is a convalescent depôt, to which detachments of invalid British soldiers serving in the Presidency district are sent; and it is garrisoned by a company of the battalion stationed at Lebong. Barracks were built on the ridge as far back as 1848, but for many years only a few invalid soldiers were quartered in them, the main body of the troops being stationed at Senchal. The latter site, however, proved unsatisfactory, owing to its excessive rainfall and depressing climate. It was accordingly decided in 1867 to enlarge the barracks at Jalāpahār, the elevation of the latter being nearly 1,000 feet lower than Senchal, the rainfall not so great, and the officers and men not so averse to it as to Senchal. The barracks now provide accommodation for 400 men; the parade ground stands 7,520 feet above sea-level. The name Jalāpahār means the hill of the burnt forest, and it is noticeable that in his account of one of the first visits ever paid to Darjeeling, General Lloyd mentions that the forests on the ridge had recently been entirely destroyed by a great forest fire.

Jorbangalā.—*See* Ghum.

Jorpokhri.—*See* Sukiāpokhri.

Kālimpong.—A village in the head-quarters subdivision, situated at a distance of 28 miles east from Darjeeling *viâ* the Rangīt road and of 32 miles *viâ* Pashok. Population (1901) 1,069. The village, which has given its name to the tract of country formerly known as Dālingkot, used to be the head-quarters of a Bhutanese district, and hence derived the name Kālimpong, the *pong* or stronghold of the Governor. According to some, the name means the Assembly of Ministers, an appropriate designation for the head-quarters of the Scotch Mission established here; but this derivation is not entertained by competent authorities.

Situation. It is picturesquely situated at an elevation of 3,933 feet above sea-level on the flanks of the Deolo hill, which towers above the village to a height of 5,590 feet. Crowning an open cultivated spur, which projects into a basin enclosed by forest-covered ridges all round, it commands a wonderful view of mountain, valley and river scenery. To the north a series of hill ranges terminates in the line of perpetual snows; to the west is the valley of the Rangīt winding among the mountains; to the south is the forest-clad Senchal spur; and to the east looking across the beautiful valley of the Rillī is ridge after ridge covered with great forests. The bazar itself lies on the saddle of a ridge in the centre of this amphitheatre of hills; and in the foreground, on the sides of the ridge sloping down to the Tīsta and Rillī, are wide stretches of cultivated fields dotted with small homesteads, which present a peaceful scene of rural comfort.

Buildings The outstanding landmark at Kālimpong is the handsome Gothic Church and tower just above the bazar, which were built as a memorial to the Revd. W. Macfarlane, the pioneer missionary of the Church of Scotland in the Darjeeling district, who died here in 1887. On the slopes of the hill near this Church are the various buildings of the Guild Mission, which is so called, because, while part of the whole Eastern Himalayan Mission of the Church of

Scotland, it is supported by the Guilds of that church. These buildings include a well-equipped hospital known as the Charteris Hospital, and a Medical Mission House, a Ladies' Mission House and School, the Scottish Universities Mission House and Training and High School, the Boys' Hostel and the Tibetan Mission House. In front of the church are the Guild Mission House and the Prince of Wales Industrial School, and below these again in the bazar are the Coronation Dispensary and the Victoria Memorial, a handsome open porch of Tibetan design which was erected in memory of Queen Victoria. The pillars and cornices were carved by Lāmas from Sikkim, and have been painted in bright colours; inside is a bronze bust of Queen Victoria standing on a pedestal from which issues the main water-supply of the village. Near the bazar on the slope of the hill is a Buddhist monastery surrounded by tall bamboo poles from which flutter long narrow strips of cloth, the prayer flags of the Buddhists. The village also contains a Hindu temple and Muhammadan mosque.

The Colonial Homes. Further along the ridge are scattered the buildings of the St. Andrew's Colonial Homes, which are intended for the education and training of poor European and Eurasian children. They were founded in 1900, under the auspices of the Church of Scotland, by a Board of Management composed of officials, tea planters, missionaries and merchants. Their object is to give the children of the domiciled community a course of training, in a healthy district and in favourable environments, such as will fit them for emigration to the colonies and for honest labour in India. It is a well-known fact that the domiciled community deteriorates in the environments of a tropical country and oriental standards; and the object of the Homes is to endeavour to break down the influence of heredity and of such environments by removing the children at an early age to surroundings which are healthier, both physically and morally, than the towns in the plains, by teaching them the dignity of labour, and by instilling principles of self-respect, self-reliance and self-help. The system adopted is that of cottage houses, each cottage holding 25 to 30 children and being in charge of two ladies from Great Britain. Special attention is paid to removing the distaste for manual labour which is characteristic of the Eurasian and poor European. All the work of the cottages is, therefore, done by the children, no servants being allowed; and they are occupied, when old enough, in farm labour, carpentering, etc. The Homes are situated on an estate of 400 acres, at a height ranging from 4,500 to 5,500 feet above sea-level; and comprise the cottages mentioned above, a central school, farm steading and workshops.

The scheme is managed by an independent committee, as the Homes are not an integral part of the Mission; but the missionaries, who initiated the scheme, superintend the local arrangements.

The Scotch Mission.

The Guild Mission itself was started at Kālimpong in 1873, the work being carried on by native teachers until an European missionary settled there in 1880. The Mission is now firmly established, and has done most useful work among the cultivators of the Government estate. Practically, the whole of the education in the tract east of the Tīsta is under its control; it has developed technical education by the establishment of weaving, carving and lace schools; and the work of the Medical Mission, which has a hospital at Kālimpong and a dispensary at Nimbong, is deservedly appreciated by the peasants. The European staff consists of an ordained missionary, an ordained Principal of the Training College, a medical missionary; a lay missionary engaged in work among the Tibetan population, 2 missionary nurses, a lady teacher for the girls' school, and 3 other missionary ladies, besides 2 ordained native pastors and 11 catechists for the 13 parishes into which the Kālimpong estate has been divided. At Kālimpong itself the Mission maintains the Charteris Hospital with 28 beds, at which compounders are trained; a training school, at which teachers for all the Mission schools in Darjeeling and Sikkim are trained; and an Anglo-Hindī school, teaching English up to the Entrance standard of the Calcutta University. Outside Kālimpong it maintains 35 vernacular schools with over 1,100 scholars, and it has also a branch mission at Gorubathān, which is the head-quarters of an ordained missionary, who is also the chaplain of the Duārs planters.

Kālimpong is a place of considerable local importance, as it is the head-quarters of the Tīsta Forest Division and of the Kālimpong Government estate, and is also the commercial centre of the tract east of the Tīsta and the entrepôt for the Tibetan trade. Besides this, it is the chief market for the agricultural produce of the neighbourhood, to which the villagers bring the produce of their fields on market days. The trade of the place has been very much stimulated by the establishment of an annual *melā* or fair at the end of November. This is the chief event of the year to the peasants of the whole neighbourhood. They observe it as a two days' holiday, and flock in to do their marketing and to enjoy the athletic sports, dancing and other amusements which form part of the fair. Merchants from more distant parts also come in from Nepāl, Sikkim and Bhutān, and even from places in Tibet a month's journey or more away, bringing various kinds of merchandise and numbers of mules which are bought up by Government for transport. An agricultural and horticultural exhibition is held at the same time, at which prizes are given for grain, poultry, cattle, mules, ponies, dairy produce, and articles of local manufacture. The fair was started in 1891 by Dr. Graham, the head of the Guild Mission at Kālimpong, and has developed year by year, till it is now one of the largest fairs in the hills. For a fuller account of Kālimpong, *see* Dr. Graham's book "On the Threshold of Three Closed Lands (1905)."

Kālimpong Government estate.—A hilly tract formerly known as Dālingkot, situated east of the Tīsta, west of the Ni-chhu and Di-chhu (Jaldhākā) rivers, and south of independent Sikkim. It was acquired from Bhutān after the campaign of 1864-65. It lies between 26° 51′ and 27° 12′ N., and 88° 28′ and 88° 53′ E., and extends over 401 square miles. Of the total area 213 square miles are occupied by reserved forests and 10 square miles by 4 tea gardens, while 178 square miles are reserved for native cultivation; five-sixths of the inhabitants are settled on the Khās Mahāls or State lands. The country is cut up by ridges of varying height and steepness, separated by narrow valleys, the principal among which run far back into the mountains. These ridges debouch into the plains at elevations ranging from 300 feet to 1,000 feet above sea-level, rising in the interior to 10,500 feet at Rishi-lā. Over a large portion of the tract the reserved forests cover the tops of the ridges and the bottoms of the valleys, while the cultivated area occupies the intervening space. The land above 5,000 feet is generally, and that above 6,000 feet almost entirely, under reserved forest, which also covers most of the area below 2,000 feet. The chief crop grown is maize, which occupies 38,000 acres or more than three-quarters of the net cropped area.

The land has been classified for revenue purposes as (1) cardamom land, held rent-free for the first three years, during which there is practically no outturn, after which it is assessed at Rs. 10 per acre; (2) terraced rice lands, paying from 8 annas to Re. 1-4 per acre, (3) unterraced cultivation, including fallows of less than 3 years' standing, paying 6 annas to 15 annas per acre; and (4) fallows of 3 years' standing and over, paying from 2 to 3 annas per acre; some lands in each of the last 3 classes are assessed at a slightly lower rate for the first few years of the settlement. The estate has been divided into 48 blocks, excluding Kālimpong bazar, each under a headman or *mandal*, who is responsible for the collection of rents, the repair of roads and certain other duties, in return for which he receives a percentage on the collections and other privileges. The total rental of the estate, as fixed by the settlement concluded in 1903, is Rs. 31,589. The chief village in the estate is Kālimpong, and there are large bazars at Pedong, on the Tibetan trade route, and at Sombāri at the end of the Chel valley, where the produce of the hill cultivators is sold to the cultivators of the Duārs. For further information regarding the estate, *see* Chapter XII and the Report on the Survey and Settlement of the Kālimpong Government Estate by Mr. C. A. Bell, I. C. S. (1905).

Kurseong.—Administrative head-quarters of the Kurseong subdivision, situated on the Lower Himālayas, 4,860 feet above sea-level and 20 miles from Darjeeling, in 26° 53′ N. and 88° 17′ E. Population (1901) 4,469. Situated on the last ridge to the south which screens Darjeeling from the plains, Kurseong, though deprived of the magnificent panorama of the snows which is seen from the former place, still commands a fine view of the Ghum ridge to the north, while the peaks of Kinchinjunga can be seen showing through the gap at Ghum between that ridge and the Senchal spur. The Ghum ridge limits the view to the north; but nature has filled up the intermediate space with a wealth of detail, which goes far to compensate Kurseong for the limited view of the snows. The Nāgri spur, starting at Jorpokhri and stretching down to the Bālasan, is particularly picturesque; and whichever way one turns, is a magnificent outlook over hill-sides studded with tea gardens and factories. But to many the real charm of Kurseong is to be found in the view to the south. Choosing a clear day and taking his stand on Eagle's Crag or the ridge towards Constantia, the spectator commands a wonderful view of the plains of India, stretching 100 miles away to the distant horizon. One feature of the landscape which cannot fail to strike the observer is the suddenness with which the hills dip down into the plains. There is in fact something incongruous in the vast plain, flat as a billiard table, on one side, and on the other a jumble of hill and valley, with spur on spur jutting out into the plains till they become lost in the dim blue of distance.

Scenery.

Looking to the south-east one sees the Tista, whose waters have come straight from the snows, flowing through dense forest at the base of the hills till it widens out into broader reaches close to Jalpaigurī. The next river looking from left to right is the river that goes under two names, being called Mahānadī by the people of the hills, and Mahānandā by those of the plains. This river joins the Bālasan some 4 miles west of Silīgurī, the combined river, retaining the name of Mahānandā, flowing to the south-west on its way to join the Ganges. The Bālasan, after taking its waters from springs in the Ghum ridge near Pulungdung, flows beneath Kurseong and comes out on the plains at Pānighātā, where it is spanned by a suspension bridge. Still further to the right flows the Mechī, which divides us from Nepāl; beyond that again are the dense jungles of the Morung district, the breeding ground of elephants; and, on a very clear day, two more Nepāl rivers may be seen,—the Maikhola and the Jokmai. A bird's eye view of the Tarai at once discloses how the hand of the cultivator has left its mark on the landscape. The primeval forest has largely been cut down, and the numerous patches in various shades of green afford a means of discriminating between tea, paddy, and jute crops; dotted about in the clearings are the tin roofs of tea factories; while the dull brown of certain patches, which appear towards the close of the winter season, testifies to the way in which large herds of cattle have done justice to the grazing. To all but those who have sought a refuge from the sweltering heat of the plains, the land invites a closer acquaintance, and one is enabled to understand something of the feelings of the Aryan invaders when they caught their first glimpes of the land that was to be their heritage.

In the 18th century Kurseong was a village in Sikkim, but about the first decade of the 19th century the Nepalese conquered and annexed the Tarai and the lower hills, including Kurseong. **History.** At the close of the Gurkha war, in 1817, the British restored the country to the Sikkimese, and, in return for this, claimed a paramount influence in Sikkim, which has survived to the present day. Finally in 1835, a strip of hill territory, about 5 or 6 miles wide, stretching from the northern frontier of the district to the hills below Pankhābāri, was ceded to the British by the Rajā of Sikkim. This tract included Kurseong, which appears to have been at that time an insignificant village, as well as a few miles of land on either side of it. In 1880 the Darjeeling-Himalayan Railway was extended to the place, bringing it into direct communication with Calcutta. From that time it began to gain popularity as a hill station; and in 1891, it was made the head-quarters of a subdivision including the Tarai and a portion of the hills.

Kurseong is a small hill station extending in three directions from the railway station, which is 4,864 feet above sea-level. The railway line winds its way northwards through a narrow bazar, **The town.** which strikes the traveller as a curious compromise between the hills and the plains. A few Tibetan *dandywālas* lounging about the road show the traveller that he is among the Himālayas; here and there a glimpse of prayer flags may be seen up a side street; while the bright complexions of the women and the Mongoloid features of the Nepalese emphasize the fact that he has arrived at the threshold of the Mongolian races. The foreign element, however, predominates. The Mārwāri from the plains with his jaunty *pagrī* is much in evidence, while the presence here and there of a Musalmān thinking of his hide business, or of a Chinaman contractor on a sleek well-kept pony still further accentuates the mixture of races. Passing by the dāk bungalow and the butchers' quarter, the bazar is left behind, and the tea-clad hill-sides again come into sight. Just below the road are the pretty little houses of the Monteviot Estate nestling along the side of the hill: 3,000 feet or more below is the meeting ground of the Bālasan and Rangbong, a vast expanse of shingle and boulders; while the Ambutiā flat looks as if parts of it were specially designed by nature for a race-course. On the other side of the Bālasan lies Mirik, an ideal site for a hill-station, and a prominent point on the range which connects the plains with Tanglu, Sandakphu and Phalūt; and beyond Mirik the Nepāl hills are visible. Continuing our course along the Cart road, we pass the new Jesuit Church on our right, and, just above the Church, St. Helen's, a Convent of the Daughters of the Cross, where there are 16 nuns and 127 children. This institution was founded in 1890, and has an orphanage for hill children, called St. Margaret's, attached to it. A little further on, at a distance of about three-quarters of a mile from the railway station, we come to the Clarendon Hotel, and beyond this lies St. Mary's Seminary, an imposing building belonging to the Jesuits, which was erected in 1888, and contains 39 theological students, the course of study extending over 4 years. Still further along, about half way to Sepoydhura, is the new building of the Industrial School which is managed by the Irish Christian Brothers.

Coming back towards the Hotel, a branch road up the hill to the left takes us to Dow Hill, which commands Kurseong on the east and shuts it out from the Tīsta valley. Dow Hill is part of the ridge which culminates on this side in the spot called Chimney and to the north in Tiger Hill, after which it merges into the Jalāpahār ridge as we approach Darjeeling. The old Military road, which rises by a fairly steep ascent from the railway station and runs along the crest of the ridge as far as Ghum, brings us at a distance of 2 miles from the station to the European schools, the Victoria Boys' School and the Dow Hill Girls' school, which occupy an local site, 6,000 feet above sea-level, overlooking the rest of Kurseong. The boys' school lies about half a mile to the south of the girls' school, and is reached by a road passing through a delightful fir forest. Both schools are Government schools and contain 190 and 120 children respectively. Sites have also been selected for a training college for teachers and a preparatory school for boys; and there is every indication that an extensive colony will soon spring up on Dow Hill. There are already, in fact, several residences on both sides of the old Military road between Kurseong bazar and the schools, and ample building sites are available. Still higher up the hill lies Chimney, at a height of about 7,200 feet, where the old Military road comes out on the other side of the ridge, overlooking the Pāglājhorā. This curious name is due to the fact that in the days when this was the regular road to Darjeeling, there was a dāk bungalow at this spot, of which only a single chimney now remains. In Dow Hill are the springs which feed the municipal water-supply and furnish the residents of Kurseong with an average of about 60,000 gallons per diem. Some scarcity is felt during April and May, but there exists no serious cause for apprehension, as there are good springs above Sepoydhura, which could readily be brought into requisition, if necessary. The rainfall on Dow Hill is also much greater than at Kurseong itself, averaging 220 inches per annum as against 165 inches at the public offices.

Descending again to the railway station we follow the course of the Pankhābāri road, which, leading westward from the station, skirts Eagle's Crag on the north side. It then slopes gently downwards to the Church, *Kachahri* and other public offices, at a distance of about a mile from the station, and runs nearly level along the crest of the ridge till it reaches the turning to Namsu, where it dips down the lower ridge to the south to Pankhābāri. Above the junction is Constantia, the house of the Subdivisional Officer, at the extremity of the Kurseong ridge overlooking the Bālasan. The Church, *Kachahri*, Amusement Club, Dispensary, Jail, and Municipal office are all situated within a compass comprising a few acres just below the Pankhābāri road.

Of late years, in view of the scarcity of available building sites in Darjeeling, public attention has been attracted to the possibilities of Kurseong as a supplementary hill station, the climate being similar to that of Darjeeling, but much milder, and the place being conveniently situated for those who dislike the higher altitude or the severe winter of Darjeeling. Within the last decade many new houses have sprung up and there has been a noticeable expansion of the town. This movement has been especially marked during the last four years since the inauguration of the Kurseong Improvement Committee and the construction, under their supervision, of the Bourdillon road, which connects the Monteviot end of the town with the Pankhābāri road and affords access to a number of fresh building sites. A Scotch Mission Church was built in 1904; in 1905 the Jesuit Church was completed; and a horse and flower show was started in the latter year, which bids fair to become a yearly institution. A new *Kachahri* was built in 1904, and a hospital and dispensary containing a European ward are shortly to be erected in a central position in the bazar. Negotiations for the building of a new club, on the piece of land known as the Recreation Ground, are also in progress, and it is hoped that its erection will supply a want which has long been felt, more particularly by visitors. Possessing as it does great natural advantages and an almost unlimited number of building sites, the place is capable of almost indefinite expansion; and it lies with the people of Kurseong themselves to advance its development and to make it a prosperous hill station.

To this sketch* of Kurseong may be added the following extract from an inspection report written by the Deputy Sanitary Commissioner in 1905:—

"The increasing importance of Kurseong has never been so prominently brought forward as this year. In spite of plague and other drawbacks its popularity has increased by leaps and bounds. One reason for this may be the new type of accommodation which is being gradually provided. Another is probably that the average climate compares favourably with Darjeeling, the normal daily average being little over sixty degrees. For many complaints this is recognized to be of some importance, while the rapid drop in the evening temperature of Darjeeling is recognized as correspondingly treacherous. There seems then, with only two hill stations in Bengal ample scope and room for the betterment of the lesser. What is necessary is a comprehensive policy with the object in view that Kurseong must be developed into a flourishing hill station and sanitarium, not rivalling but complementary to Darjeeling".

Kurseong subdivision.—Southern subdivision of the district, lying between 26° 31′ and 27° 0′ N., and 88° 7′ and 88° 31′ E., and extending over 438 square miles. Its population was 115,731 in 1901 against 117,642 in 1891, and was contained in one town Kurseong, its head-quarters, and 388 villages, the density being 264 persons to the square mile. The subdivision contains two distinct tracts, the northern resembling the adjoining portion of the Darjeeling subdivision, while the Silīgurī thāna lies in the plains. After Kurseong, the most important places in the subdivision are Silīgurī, the junction of the Eastern Bengal State and

* I am indebted to Mr. H. R. T. S. Perrott, I.C.S., for the description of Kurseong.

Darjeeling-Himalayan Railways, Tindhāria, where the workshops of the latter railway are situated, and the large marts of Mātighara and Naksalbāri in the Tarai. The subdivision was formed in 1891 out of the southern portion of the head-quarters subdivision and the old Tarai subdivision. The latter had been in existence since 1864, its head-quarters being at first at Hānsquar (Hānskhāwā) near Phānsidewā, where there are traces of the old *Kachahri* buildings on a hill about 150 feet above the plain. In 1880 the head-quarters were transferred to Silīgurī, which had already been brought into direct communication with Calcuttā by the extension of the Northern Bengal State Railway, and which was the terminus of the Darjeeling-Himalayan Railway then under construction. Though that place has ceased to be the administrative centre of a subdivision, it is still the head-quarters of a Deputy Magistrate, who is in charge, under the Subdivisional Officer, of the judicial and revenue administration of the Tarai.

Lebong.—A spur situated just below the town of Darjeeling at an elevation of 5,970 feet above sea level. Lebong, or as it is called by the natives Alibong, is a Lepchā name, meaning "the tongue-shaped spur," which has been suggested by the fact that it shoots out like a tongue from the ridge on which Darjeeling stands. Lebong, with its more equable climate, was one of the first places in the district to attract European settlers. It was found that Lebong was about 10 degrees warmer than the latter place, and enjoyed more sunshine. Locations were accordingly taken up on it at the same time as at Darjeeling, and the cultivation of tea was begun on its slopes. A large portion of the spur has now been acquired for cantonments; barracks and a parade ground have been laid out; and a battalion of British infantry is stationed there.

Mangpu.—A village in the head-quarters subdivision, situated 18 miles south-east of Darjeeling and 5 miles east of Sarail. It is the head-quarters of the cinchona plantations of Bengal, of which the following account is taken, with some abbreviations, from an interesting article on "Cinchona and Quinine in the Eastern Himalayas" published in the Kālimpong Saint Andrew's Colonial Homes Magazine, October 1906.

The home of cinchona is in South America, on the slopes of the Andes, at an elevation of from 3,000 to 8,000 feet. As it first came from Peru, its bark was known as "Peruvian bark." The name "Cinchona" was given because the Countess of Chinchon, the wife of the Spanish Viceroy of Peru, was cured of fever by it as early as 1638. Another early name was "Jesuits' bark" either because it was disseminated by that Order in Europe or because it was discovered by a Jesuit missionary who is said to have benefited from its administration by the local Indians. About 1655 its healing virtue was made known in England. An apothecary named Robert Talbor by its means cured Charles II of a tertian fever in 1679, and subsequently treated successfully the Dauphin and others in France. This led Louis XIV to buy from him the description of the treatment, which was published and afterwards translated into English. In 1677 "Peruvian bark" appeared as a regular medicine in the London Pharmacopæia, and in 1742 Linnæus established the genus cinchona. For two centuries all the bark was got from South America, where collectors sought out the trees scattered in the dense primeval forests, felled and barked them. But this was a wasteful method, and with an increased demand there was the danger of exhausting the supply, and efforts were therefore made to cultivate the plant in other countries. The pioneer attempt was made in Algeria in 1849; and in 1854 the Dutch introduced it to Java and laid the basis of the present industry in that island. The large consumption of bark in India then led the Government to take effectual steps to introduce the cultivation of cinchona into India, and accordingly in 1857 Mr. Clements Markham was sent to obtain a supply of young trees from South America. In 1861 a consignment was planted on the Nilgiri Hills, where there is still a large Government Plantation and Factory. In 1862 an experiment was also made by Government, near Darjeeling, under Dr. Anderson, of the Botanical Gardens, Calcutta. Later on a large tract of land in the neighbourhood was set apart for the purpose, and this is the plantation of which Mangpu is now the centre. Since that time much has been done by careful cultivation, selection, hybridizing, grafting, etc., to improve the medical qualities of the bark, and an equal advance has been made in the process of manufacture.

The cinchona is, at least as regards the better kinds, a delicate plant, and requires most careful treatment. The extremely fine seed is sown in beds; the tiny plants are transplanted and re-transplanted in shaded nurseries before being put out in regular rows (4 to 6 feet apart) on the mountain sides. Much labour has to be expended on hoeing, and manuring is resorted to where feasible. Bark harvesting, in progress throughout the whole year, usually commences when the trees are three years old. At this age, thinning is necessary where overcrowding exists, and individuals are uprooted as they show signs of unhealthiness. Every year the whole plantations are thus searched over and trees removed where necessary. To collect bark, the trees are uprooted and divided into three parts - roots, stem, and branch—each of these contaning alakaloids in different proportion. All the bark is srcaped off with blunt knives, and the three kinds, dried in open air sheds, are stored separately, whence they are taken to the quinine factory for the extraction of their alkaloids.

Many different species of cinchona are grown, but the chief varieties cultivated at Mangpu and Munsong are *Ledgeriana*, two Hybrids, and the robust but less rich *Succirubra*. The first of these was discovered in 1865 by Charles Ledger, a trader in wood and bark, living in Peru and Bolivia. A rich species was known by him to exist, but in districts inaccesssible to Europeans by reason of the jealousy of native bark traders. He therefore employed a trusted Indian servant, who, after passing through much ill-treatment from which he shortly afterwards died, succeeded in bringing to his master a small quantity of seed. This was offered by Ledger to the Indian Government, but being declined by them, was sold, half to the Dutch Government for the Java plantations and half to Mr. Money, a planter on the Nilgiris. Some trees from this seed, analysed at maturity by the Dutch, were found to yield bark containing the equivalent of 9·97 per cent. sulphate of quinine, whereas formerly the cinchona cultivated in Java yielded but a quarter of this quantity.

What makes cinchona so valuable is that it contains those alkaloids which have a curative effect in malarial fever as well as other medicinal properties. There are many such alkaloids present, but the chief are quinine (from the Peruvian-Indian word "Kina", meaning bark), cinchonidine, cinchonine and quinidine. As quinine is the most important of all, we shall confine ourselves to a description of its preparation. A speciality of the Government Factory has been cinchona febrifuge or mixed cinchona alkaloids. It contains all the febrifuge alkaloids of *Succirubra*, and as it can be made more cheaply than quinine and is almost as effective, it has proved a great boon, specially at a time when the price of quinine was prohibitive to poor people.

The factory at Mangpu is a busy centre, and one is struck with the amount of the plant needed for the various processes. Commodious store-houses are required for the reception of the dried bark, which is first of all ground into the finest of powder by powerful disintegrators, so fine that it passes through a silken screen which contains 14,400 holes to the square inch. This powdered bark is then conveyed to the digesters (large cylindrical tanks containing steam coils and "stirrers") in which it is mixed with water and caustic soda, the latter setting free the alkaloids. Oil is added to dissolve and absorb the alkaloids. The bark sediment and water fall to the bottom of the tank, and the alkaloid-charged oil is run off and put into the row of tanks (separators) opposite, fitted with stirrers to mix it with sulphuric acid, which in its turn takes up the alkaloids from the oil and converts them into sulphate of quinine, etc. The oil, now minus the alkaloid, is run off, to be used again, while the acidulated liquor containing the alkaloids, is conveyed to large lead lined tilting pots and there mixed with sufficient caustic soda to neutralize the excess of acid. The quinine crystallises out, after which the crude crystals are filtered, re-dissolved in hot water, and treated for the removal of a colouring matter and impurities. The liquid is again filtered, and allowed to cool. The quinine crystallizes out, the crystals are drained in a centrifugal machine, dried in a warm room, packed and sent off for distribution to the Botanical Gardens, Calcutta.

The laboratory becomes every year an increasingly important element in the organization of the Cinchona Plantations. Here the bark of selected trees is analysed with a view to the discovery of those containing the highest percentage of alkaloids, and these alone are used for seed. In the laboratory, too, the processes which we have described on a large scale are gone through in miniature, and so a check is kept on the manufacture.

The quinine manufactured at Mangpu is used exclusively for the Government's own requirements and is distributed to the various hospitals and dispensaries. A considerable profit has accrued from the manufacture. A few years ago the experiment was tried of selling it in pice-packets through the Post offices so that it might be available for the poorest peasant and in the most remote districts. The packets containing 7 grains each are made up by prisoners in the Alipore Central Jail, Calcutta. Last year 4,500 lbs. were issued for this purpose. The total output of the Mangpu factory in 1905-06 was 15,777 lbs., and at present new machinery is being added to bring up the output to 20,000 lbs. a year.

Apart from the presence of the cinchona itself, Mangpu is a most attractive spot. It is beautifully situated, with a view of the snows as well as of the plains of India.

Morung.—A name formerly applied to the southern portion of the district from the Mechī on the west to the Tīsta on the east, which is now known as the Tarai. It was annexed from Sikkim in 1850 in consequence of the seizure and detention of Dr. Campbell, the Superintendent of Darjeeling, and Sir Joseph Hooker, while travelling peaceably through the Rājā's territories. At that time, the upper portion of this tract, lying immediately at the base of the mountains was chiefly covered with forest and jungle, and was inhabited by two aboriginal tribes, the Meches and Dhimāls, who did not suffer from its unhealthy climate, but got ill on leaving it for the open plains. The lower portion was more open and cleared, was principally cultivated with rice, and was chiefly inhabited by Koches or Rājbansis.

Pedong.—A village to the north-east of the district situated 13 miles north-east of Kālimpong at a height of 4,760 feet above sea-level. It contains a dispensary, a Buddhist monastery, an office for the registration of trade with Sikkim and Tibet, and a rest-house for travellers; and it is a commercial centre of some local importance. It is the head-quarters of the Roman Catholic Mission to Tibet, which has here a presbytery, an orphanage and a chapel erected in 1883. An account of the establishment of this mission will be found at the end of Chapter III. A grant of land has been given to the mission by Government in the neighbouring village of Maria-bastī, where only Christians are allowed to settle, and where a large stone church has been built. The name Pedong means "the incense-tree clearing" and is derived from the *sāl* (*Shorea robusta*) trees which once grew there, the resin of which is used as incense in Buddhist temples.

Phalūt.—One of the loftiest peaks in the Singālilā range situated 50 miles north-west of Darjeeling at a height of 11,811 feet above sea-level. The name Phalūt is a corruption of the Lepchā Fak-lut or "the peeled summit," an appellation derived from the bare treeless slopes, which offer a great contrast to the forest-clad ranges below. The peak is also called Phalalum (*Pha-li-lung*), a Tibetan mispronunciation of the Lepchā name. A magnificent view of the snowy range is obtained from Phalūt, and in particular of Kinchinjunga, which is only 30 miles distant. Mount Everest is however scarcely visible, as it is hidden by Peak No. XIII, a mountain resembling in shape an enormous armchair. The general panorama resembles that described in the article on Sandakphu, except that Kinchinjunga and the mountains on either side loom much larger, this being the nearest view of them which can be obtained in the district. The range round Phalūt itself has not the same wealth of vegetation as at Sandakphu, for it is bare of forest and even of shrubs, except rhododendrons, and consists of grassy undulations studded with masses of gneiss rock.

Rangarun.—A forest stretching along the northern flank of the Senchal and the western flank of the Takdāh range, in the midst of which is a bungalow under the control of the Deputy

Commissioner, situated 7½ miles from Darjeeling at a height of 5,700 feet above sea-level. A Botanical Garden was established here in 1876 with the object of introducing trees of the coniferae order, cedars, firs, pines, larches, spruces and cypresses, and of developing the culture of various species of rhododendrons and other flowering trees and shrubs. But the site was found to be too exposed and the plantations suffered from hail-storms, so that it was abandoned in 1878, and in its place a Botanical Garden was established in Darjeeling. A fine view can be obtained from the bungalow of the Darjeeling ridge across the deep Rangnu valley with the snowy range in the background, while the great creeper-bound and moss-covered trees all round present a magnificent picture of forest scenery which cannot be equalled in the neighbourhood of Darjeeling.

The name Rangarun is also applied to the forests on the eastern slope of the Takdāh ridge looking down on the Tīsta valley. It is a Lepchā name, which, like the name Rangli Rangliot, is connected with a legend concerning the rising of the Tīsta high above its bed. According to the legend, the Rangīt quarrelled with his spouse, the Tīsta, and, parting from her, carried his waters high up the hill-side to Rangarun and Rangli Rangliot. Then, fearing that the world might be inundated, he returned and rejoined the Tīsta, and the two rivers have flowed on in peaceful union ever since. The two places Rangarun and Rangli Rangliot, which mean "the turning of the great river" and "the brimful great river," are said to mark the spot up to which the water rose and then receded. The legend probably preserves the memory of some great landslip, which dammed up the river and forced it to rise high up the valley.

Rikyisum, or as it is more usually spelt Rissisoom or Rissoom, is the name of a forest to the north-east of Kālimpong. A bungalow, which is under the control of the Deputy Commissioner, has been built here, in a spot of great natural beauty, at a distance of 12 miles from Kālimpong and at a height of 6,410 feet above sea-level. It is set in the midst of some of the finest forest scenery to be seen in the district, and looks out upon a splendid panorama of the snowy range, the summits of which rise above a series of deep intervening valleys. The name means "the three ridges," as it marks the place where three spurs diverge.

Sandakphu.—The loftiest peak in the district, situated 37 miles to the north-west of Darjeeling on the Singālilā range, and rising to a height of 11,929 feet above sea-level. The name Sandakphu means the hill of the poison plant, and has been given to this peak, because the deadly aconite plant grows thickly along the slopes for about 2,000 feet below the summit. For this reason, cattle not habituated to its use have to be muzzled in order to prevent them eating the succulent but poisonous plant. Sandakphu commands the finest view of the Himālayas to be obtained in the district. In the foreground is a great basin set in the midst of the hills, the slopes of which are covered by masses of rhododendrons and fragrant pine forests. In the background is a continuous barrier of snowy mountains, the most prominent of which is Kinchinjunga towering up in gigantic height and breadth, with its attendant peaks, Kabru, Jāno and Pandim, clustering closely round it. Mount Kabru is here foreshortened and does not present the same graceful outline as when seen from Darjeeling, but Jāno rises far higher above the spectator with its lofty peak standing up like a great icy horn. Far off to the west, at a distance of 100 miles from the observer, the graceful peak of Everest is seen among a group of other snowy mountains, rising from behind the crest of Peak No. XIII, which in shape curiously resembles a great arm-chair of snow; and further to the west, towards Nepāl, there is a wonderful square mass of mountains looking like a wall of snow. The contrast between Everest and Kinchinjunga is very marked. The latter is remarkable for its imposing bulk and massive proportions, while Everest, soaring above a series of valleys and ridges, is more graceful and majestic. The space between the two is occupied by snowy ranges of smaller proportions; and eastward, beyond Kinchinjunga, are visible, first the Narsingh group, and then the Dongkya and Cholā ranges on the Tibet frontier, with Chumalhāri lifting up its head in the rear. The whole snowy range of Bhutān, Sikkim and Nepāl, about 200 miles in length, is visible, but the panorama is completely dominated by the Kinchinjunga and Everest groups.

Sarail.—A village in the head-quarters sub-division, situated at a height of 5,615 feet above sea-level, 5 miles west of Mangpu. It contains an inspection bungalow and is the most convenient place for a visit to the Cinchona Plantations at Mangpu. *See* MANGPU.

Senchal.— A mountain situated 6 miles to the south-east of Darjeeling at a height of 8,163 feet above sea-level. Just above Senchal a peak known as Tiger Hill rises to a height of 8,515 feet and further to the south-east is West Senchal at an elevation of 8,600 feet. The name Senchal means "the hill of damp and mist," and the appellation is well deserved, for the ridge catches the full force of the monsoon driving up from the Bay of Bengal, and for the greater part of the year, is drenched by rain or enveloped in mist. In 1844, in the early days of Darjeeling, it was chosen for the site of a cantonment, and it was occupied by troops for over 20 years. It was at last recognized, however, that the site was unsatisfactory, owing to the excessive rainfall and depressing climate of this dismal exposed ridge; and it was accordingly decided in 1867 to place the barracks on Jalāpahār, the elevation being nearly 1,000 feet less than that of Senchal, the rainfall not so great, and the officers and men not so averse to it as to Senchal. The barracks on Senchal were then demolished, and nothing is now left of them except a few solitary chimneys, some deserted ruins, and neglected paths overgrown with jungle and grass. According to local tradition, the cold, mists and loneliness of the place drove many of the soldiers to commit suicide; but the number of suicides appears to have been exaggerated, if we may judge by the number of graves in a little cemetery on the side of the road leading along the eastern side of the Jalāpahār ridge to Ghum. Here a large Irish cross of stone bears a tablet with an inscription saying that it was erected in memory of the officers, non-commissioned officers and men who died at Senchal during the years 1844 to 1865 and whose mortal remains rest there. There are, however, only 14 nameless and neglected graves in the little cemetery.

The ascent to Senchal winds along the western flank of the hill passing through magnificent forests of oak, magnolia and rhododendron, with thickets of wild raspberries on either side and sparkling cascades tumbling down the hill. The summit commands a fine view of the plains of India, of the mountainous spurs sinking to the level plains, and of the courses of several great rivers, the Tīsta, Bālasan, Mahānadī and Mechī. To the north the prospect is far grander. In the foreground is the great valley of the Rangnu, 4 miles across and fully 4,000 feet deep, which is formed on one side by the Darjeeling ridge, bare of forest and scarred by land-slips, and on the other by the forest-clad Takdāh range. Beyond this valley is a line of snowy mountains stretching across the horizon. In the centre is the massive bulk of Kinchinjunga 45 miles distant (28,146 feet), flanked on the right by the tent-like Kabru (24,015 feet) and the more graceful Jāno (25,300 feet), and on the left by the sharp conical peaks of Pandim (22,020 feet) and Narsingh (18,145 feet). To the north-east Chumalhāri (23,933 feet) is seen, at a distance of 82 miles, rearing its head as a great rounded mass over the snowy Cholā range, though it really lies 40 miles beyond. To the north, again, at a distance of over 100 miles, three snowy peaks just appear above the black Singalīlā range, the middle one of which is Mount Everest. The great distance, however, makes it appear of miniature dimensions, and as a peak it is completely dwarfed by the larger mountains of Sikkim.

On the north-western slope of Senchal are the springs and tanks from which Darjeeling draws its water-supply; and on the top of the ridge golf links have been laid out, which are among the highest in the world.

Silīgurī.—Village in the south of the Kurseong subdivision, situated near the left bank of the Mahānadī in 26° 43′ N. and 88° 26′ E. Population (1901) 784. Silīgurī is the northern terminus of the Eastern Bengal State Railway, where it is joined by the Darjeeling-Himalayan Railway. It is also the terminus of the cart road from Kālimpong and Sikkim, and it thus focuses the local trade. Several jute firms are established here, and in addition to the permanent shops there is a biweekly Government market. It is the head-quarters of a Deputy Magistrate, who disposes of the criminal work of the Tarai and manages the large Government estate; he was formerly stationed at Hānsquar (Hānskhāwā) near Phānsidewā, but his head-quarters were removed to Silīgurī in 1881 on the extension of the railway to that place. It also contains a small sub-jail and post-office, dāk bungalow, inspection house, police station, and a dispensary with 20 beds, all situated near the railway station. The village is situated on fairly high ground, and its name means "the stony site," presumably because the bed of the Mahānadi close by is a mass of broken stone brought down from the hills.

Sonāda.—A village in the head-quarters subdivision, situated on the railway at a height of 6,552 feet, 10 miles to the south of Darjeeling. Half a mile below the railway station is the Victoria Brewery, occupying some old barracks, which were formerly used as a halting-place by troops marching up to Darjeeling. A spur about a mile to the west is known as Hope Town, where it was once hoped that a European settlement would spring up. The ground was cleared in 1856, and laid out in building sites, the idea being that a small colony of European pensioners and farmers would be established there; and it was proposed to erect a church, school-house, dispensary and bazar. A church was erected, but the place failed to attract settlers, and much of the land was bought up by planters. Latterly, however, with the facilities afforded by the railway, a few houses have sprung up, and there appear to be better prospects of developing the place. The name Sonāda, which means "the bear's den," is suggestive of the number of wild animals which used to haunt the surrounding forest. It was also formerly known as Pacheem, and was a regular halting place for travellers proceeding by road to Darjeeling.

Sukiāpokhri.—A village in the head-quarters subdivision, situated 11 miles to the south-west of Darjeeling on the road leading from Ghum to Phalūt. It is a large and prosperous bazar, under the management of the Deputy Commissioner, which attracts a considerable amount of the trade from Nepāl. The village is the head-quarters of the Nepāl Mission, which was established there in 1896 with the object of spreading Christianity among the Nepalese who come in large numbers to the market held there every Friday. The Mission maintains a dispensary located in a stone building in the bazar, which also contains a roomy mission hall. It is supported by voluntary donations, the missionaries providing for their own personal expenses and drawing nothing from the Mission funds. About a mile above Sukhiāpokhri is the Jorpokhri bungalow, picturesquely situated in the midst of forest at a height of 7,400 feet above sea-level; and 3 miles to the west is Simāna-bastī, a bazar situated at the very edge of Nepāl and containing some houses actually in Nepalese territory.

Tanglu.—A peak on the Singalīlā range, 10,074 feet high, situated 23 miles west of Darjeeling. It is more generally known as Tumling among the natives of the district. A bungalow on the summit, close to the frontier between British India and Nepāl, commands a fine view of the Nepalese valleys and of the plains of North Bengal, with the snow-fed Tīsta on the east, the Kosi on the west and a number of smaller rivers between them. The following account of the view of the snowy range from Tanglu is taken from Sir Joseph Hooker's Himalayan Journals:—"From the summit of Tonglu I enjoyed the view I had so long desired of the snowy Himalaya from north-east to north-west; Sikkim being on the right, Nepal on the left, and the plains of India to the southward. In the early morning the transparency of the atmosphere renders this view one of astonishing grandeur. Kinchinjunga bore nearly due north, a dazzling mass of snowy peaks, intersected by blue glaciers, which gleamed in the slanting rays of the rising sun, like aquamarines set in frosted silver. From this the sweep of snowed mountains to the eastward was almost continuous as far as Chola (bearing east-north east), following a curve of 150 miles, and enclosing the whole of the northern part of Sikkim, which appeared a billowy mass of forest-clad mountains. On the north-east horizon rose the Donkia mountain (23,176 feet) and Chumulari (23,929). Though both were much more distant than the snowy ranges, being respectively 80 and 90 miles off, they reared their gigantic heads higher, seeming what they really were, by far the loftiest peaks next to Kinchinjunga; and the perspective of snow is so deceptive that though 40 to 60 miles beyond, they appeared as though almost in the same line with the ridges they overtopped. Beyond Junnoo, one of the western peaks of Kinchinjunga, no continuous snowy chain was visible; the Hímalaya seemed suddenly to decline into black and rugged peaks, till in the far north-west it rose again in a white mountain mass of stupendous elevation at 80 miles distance. To the west, the black ridge of Sakkiazung, bristling with pines, cut off the view of Nepal; but south-west, the Myong valley could be traced to its junction with the Tambur about 30 miles off: beyond which to the south-west and south, low hills belonging to the outer ranges of Nepal rose on the distant horizon, 70 or 80 miles off. South and south-east, Sinchul and the Goong range of Sikkim intercepted the view of the plains of India, of which I had a distant peep to the south-west only."

Tindhāria.—Village in the Kurseong subdivision, situated on the Darjeeling-Himalayan Railway, 31 miles from Darjeeling, at a height of 2,822 feet above sea-level. The place contains the railway workshops, a railway hospital and club; and the population consists almost entirely of railway employés and their families. It is situated on the flanks of the outer Himalayan range and commands a fine view of the lower foot hills and the plains.

Tīsta Bridge.—A small bazar situated, at a height of 710 feet above sea-level, on the Tīsta valley road at a distance of 19 miles north-east of Darjeeling *viâ* the Rangīt road. The distance is however 22 miles, if the Pashok road is followed. The bazar is of some commercial importance as it stands at the junction of several roads, the Tīsta Valley road leading to the plains, the Kālimpong road running up to Sikkim, Tibet and Bhutān, and the Pashok road, passing through dense forest and prosperous tea gardens to Ghum and Darjeeling. It is placed in a scene of great natural beauty on the western bank of the Tīsta, which is spanned by a light and narrow suspension bridge 300 feet long. The river here rushes down to the plains through high banks, above which are lofty mountain slopes wooded to their summit. At a distance of about 3 miles to the north is one of the most picturesque places in the district, a deep gorge surrounded by forests, in the midst of which the clear stream of the Rangīt joins the turbid waters of the Tīsta.

APPENDIX.

PLACE NAMES IN DARJEELING.

THIS Appendix contains a list of the names of various places, rivers and mountains in Darjeeling, and of the peaks and passes visible from it, with their meanings. It has been compiled mainly from an article written by Colonel Waddell and published in the Journal of the Asiatic Society of Bengal (Vol. LX, Part I, 1891). In that article Colonel Waddell very lucidly shows how remarkably descriptive are these native names. "The oldest names," he writes, "are found to be of Lepchā origin. The Lepchās from their wild forest life are 'born' naturalists, possessing a name for nearly every natural product, animal or vegetable, whether of economic value or not. Hence they readily gave discriminating names to the chief mountains, rivers and sites in their neighbourhood. A few of these old names still survive in places where the Lepchās no longer are present. The Bhotiyas, on settling in Sikhim, bestowed their own names on many of the already named sites, partly perhaps from the fact that the meaning of the Lepchā name was not evident, and partly to express their contempt for the Lepchās. Thus, many of the hills and rivers possess two names. And since the influx of Nepālis a third synonym in the Parbatiyā dialect of Hindī has been added in several instances for rivers, mountains and already named sites . . . . . . The great majority of the names are given by illiterate persons, so that grammatical accuracy is not always to be expected. The names, as to their meaning, may be generally classed as descriptive; a few are mythological and religious, but these are chiefly confined to monastery names; and the personal designations perpetuating the names of the founders of villages are found almost exclusively amongst the Pahāriya settlements. The descriptive names predominate, and these usually well express some very obvious physical feature of the site or river, *e.g.*, of rivers, an especial tortuosity, steepness, impetuosity, shallowness or otherwise of a course or channel; of mountains, their shape, appearance, etc.; of village sites, the stony, precipitous, meadow-like character, quality of soil, jungle-product, conspicuous tree, etc."

The river names are generally of Lepchā origin, the Bhotiās merely substituting for the Lepchā suffix *ung, i.e.*, water, the suffix *chhu* which has the same meaning; while the Nepālis substitute *khola* (literally a valley) or *nadī*, the ordinary Hindī name for a river.; in the case of hill streams, however, the Nepāli name is *jhorā*, a word derived from a Sanskrit root meaning "to lay waste" which is admirably descriptive of these mountain torrents. These Lepchā river names frequently have the prefix *rang* which conveys the sense of extension or length. The mountain names are mainly of Lepchā or Bhotiā origin, the former calling a mountain *hlo* or *chu* and the latter *ri* (*e.g.*, Chumulhāri); while the Nepālis have no individual names for the various snowy peaks The names of the mountain passes again are all of Tibetan origin, owing to the fact that while the Lepchās have kept to the lower hills and valleys, the Tibetans who settled in Darjeeling, Nepāl and Sikkim frequent the cooler heights and have always kept up a commercial and religious intercourse with the table-land of Tibet; the Tibetan term *lā* or pass is also frequently applied to the mountain itself. The names of places commonly have some suffix merely meaning a place, village, etc., thus *bong*, called *pung* by the Limbus, is the Lepchā word for a resting place, *kyong* for a village, and *tām* for a level spot. The ending *bong* again is the Tibetan for a stump or foundation, and is used for a residential village; *ling* merely means a place, *sing* a field, *gang* a ridge and *lung* a valley, while the suffix *tong* is frequently a corruption of *thāng* or meadow.

To indicate the origin of the various names in the following list, the symbol L. has been used for Lepchā, T. for Tibetan or Bhotiā, N. for Nepāli or Parbatiyā Hindī, and B. for Bengali.

Ambiokh (T).—The place below the demon's shrine.

Ambutiā (N).—The place of mango trees.

Badamtam (L).—The bank of the *padam* bamboo, the giant bamboo which furnishes the Lepchās with their milk jugs, water-vessels, etc. There was formerly a forest of these bamboos at the place known by this name.

Bāghdogrā (B).—The place of roaring tigers, a village in the Tarai formerly much infested by tigers.

Bālasan (B).—The river of golden sand, a Bengali name suggested by this river's wide bed of yellowish sand. Baluāsan, the Nepāli form of the name, has the same meaning.

Bātāsia (N).—The windy site.

Chiabhanjan (N).—The mushroom pass, a name suggested by the number of mushrooms growing by this *bhanjan* or pass between the hills.

Cholā (T).—The lordly pass, said to be so called on account of its height and difficult approach. According to another derivation, it means the lake pass, a name derived from its chain of lakelets.

Chongtong (L).—The arrow-headed place, a name given to a site at the junction of two streams.

Chumulhāri (T).—The mountain of our lady goddess.

Chunābati (N).—The lime-kiln.

Dāling (T).—The place of the arrow, so called because the hill on which the old Bhutanese fort stands is sub-conical, resembling an arrow-head in shape. The name is also explained as meaning the stony site, as the fort is perched on the precipitous edge of some gneiss rocks.

Darjeeling (T).—The place of the *dorje* or mystic thunderbolt. *See* the article on Darjeeling.

Dhajiā (N).—The ridge of prayer flags.

Dichhu (T).—The old name for the Jaldhākā. It is merely a word meaning water, *di* being the Bodo word and *chhu* the Bhotiā word for water.

Dongkya-lā (T).—The frozen yak pass. A herd of wild yaks in attempting to spend the night in this pass (18,100 feet high) were frozen to death.

Gayabāri (N).—The cow-shed. The word may also be a corruption of Gehunbāri or the place of wheat; there are numerous wheat fields in the neighbourhood.

Ghum (N).—The name of the native covering made of leaves and bamboo strips, the shape of which resembles a Λ. The position of the village of Ghum, which occupies a narrow ridge on the summit of two steep slopes, has suggested this name. Another and more probable explanation is that the name means merely the bend of the hill, which is very distinct at this place.

Ghumti (N).—The turn of the road.

Gīdhapahār (N).—The vulture's hill. Here great numbers of vultures infest the rocky cliffs overlooking the plain.

Ging (T).—The stretched-out slope.

Gok (L).—A name meaning narrow and difficult of access; an old military post on a narrow promontory between the Rammān and the Great and Little Rangīt rivers.

Gorubathān (N).—The grazing station or the herd of cattle.

Jalāpahār (N).—The burnt hill. This name is an allusion to the bareness of the hill and its slopes. The Lepchās have a far more picturesque name for it—Kung-gol-hlo or the fallen-tree hill. "The tortuous spurs," writes Colonel Waddell, "running down from the steep foreshortened ridge of Jalāpahār, represent the torn-up roots of the prostrate tree: the trunk is the ridge extending to Darjeeling; and the two main branches are the spurs of Birch Hill and Lebong, from which extend the innumerable ramifications of smaller spurs that form the branchlets."*

Jaldhākā.—The hidden water, so called because in one place the stream sinks down and disappears for some distance, flowing underneath the porous gravel and detritus.

Jāno (T).—The mountain of massed rainbows. The Tibetan name appears to be Ja-ö Pung-ri bearing this meaning, and it seems likely that this name was abbreviated to Ja-ö, and thence corrupted to Jāno.

Jelep-lā (T).—The lovely level pass, so called because it is the easiest and most level of all the passes between Tibet and Sikkim.

Jorbangalā (N).—A pair of bungalows. Formerly there were only two bungalows above what is now a thriving bazar.

Jorpokhri (N).—A pair of pools.

Kabru (T).—In the Manual of the Sikkim-Bhutia Language by Graham Sandberg this mountain is referred to as Kyab-ru or the horn of protection. The name is more properly, however, Kabur, which is possibly a corruption of Kangbur or the swelling of snow; it might also mean the white swelling (*kar-bur*), but in Tibetan the adjective is not generally placed before the noun in this way. According to local native authorities, the name means simply camphor, but it is not clear why such a name should be given. It is also said that the name applies to a peak close to Kinchinjunga on the south-east, and not to the peak known to Europeans as Kabru, and that the name of the latter is really Pahung Ri—a designation now given, in the corrupted form of Powhunri, to a mountain near the Dongkya-lā.

Kāgjhorā (N).—The crows' stream, a name given to a mountain rivulet in the station of Darjeeling. The name is said to be due to the fact that a large number of crows used to crowd round the municipal rubbish heap, which was formerly situated close to this stream.

Kālapokhri (N).—The black pool.

Kālijhorā (N).—The black stream.

Kālimpong (T).—The stockade (*pong*) of the king's minister (*Kālön*). The place was formerly the head-quarters of a Bhutanese governor.

Kinchinjhau (T).—The great bearded peak of snow, a name referring to the monster icicles of this high mountain.

---

* Among the Himalayas.

Kinchinjunga (T).—A corruption of Kangchhendzönga, meaning the five treasure houses of the great snows (*Kang* snow, *chhen* great, *dzö* treasury, *nga* five), a name given to the mountain with reference to its five peaks. The highest peak, which is lit up by the glow of the rising and setting sun, is the treasury of gold; the southern peak, which remains in cold grey shade till it is silvered by the rising sun, is the treasury of silver; and the remaining peaks are the treasuries of gems, grain and holy books, which the Tibetans regard as the most valuable of possessions. The Lepchās, however, call the mountain Konglo Chu, *i.e.*, the highest curtain of the snows.

Kurseong (L).—It has been suggested that this name is a corruption of *Kurson-rip*, the small white orchid, which grows plentifully round Kurseong and that it means "the place of white orchids." Another suggestion is that it refers to a cane which used to grow there in rich profusion, and which the Lepchās in their "Rong-ring," as they term their own language, call *kur*, and that 'seong' is a corruption of *sheang*, a stick. There are still a few of these canes to be found in the forest behind Eagle's Crag.

Laba (T).—The wind-swept site.

Lebong (L).—The tongue-like spur, a corruption of *ali*, a tongue, and *abong*, a mouth.

Lepchā Jagat (N).—The Lepchā toll-bar. A place near the frontier where the Lepchās once levied toll on the Nepāli imports into Sikkim.

Lopchu (L).—This place is referred to as Lopchok or the cool stone by Colonel Waddell in his book "Among the Himalayas". It may be possible that the word is a corruption of *lap-tso*, *i.e.*, stones set up as sign-posts to show the way; the Lepchās and Bhotiās place leaves on such stones in the belief that this will prevent them getting tired. The name is also sometimes pronounced Lukchok, and may perhaps mean the pasture land of sheep.

Mahaldi (L).—The bent-going river. This river, as seen from the hills, takes a sudden bend to the right on reaching the plains, and has apparently been given this name in contrast to the Rang-nyu or straight river, the Lepchā name for the other great effluent of Sikkim, the Tīsta.

Mahaldiram (L).—The fountain-head or source of the Mahaldi, a name given to the ridge near Kurseong because that river rises below it.

Mahānadī (B).—The great river, the Bengali corruption of Mahaldi.

Manibhanjan (N).—The dip (*bhanjan*) between the hills by the *mani* (a Buddhist monument).

Mātighara (N).—The mud house. The first mud house met with at the foot of the hills; in the hills the houses are made of wood or stone.

Mechī (B).—The river of the Meches, a name suggested by this river being the western boundary of the aboriginal Meches.

Minchu (T).—The mineral springs.

Mirik (L).—The burnt hill, *mi-rek* being the Lepchā name for the burning of jungle.

Nāgri (L).—A corruption of *nak*, straight, and *grī*, a high stockaded fort.

Naksal (T).—The forest hunting grove.

Narsingh (T).—The uplifted nose (properly Narseng), a designation descriptive of the mountain as seen from Lower Sikkim.

Paglājhorā (N).—The mad torrent, a name due to the violent outbursts of this mountain stream, which carries off the drainage of a vast ridge, and is therefore liable to sudden freshets.

Pandim (L).—The king's minister. The high peak next to Kinchinjunga is so called because it is considered to be the attendant of the god of the mountain. It has also been suggested that the name may mean the conical hill with a point like a needle.

Pānighatā (N).—The water-mill, the place where grain is ground by water-mills.

Pankasāri (T).—The pasture land of *kasha* grass.

Pashok (L).—The jungle. Here the dense subtropical jungle or forest of the Tīsta valley commences.

Pedong (T).—The halting place at the Po tree, *i.e.*, the *sāl* or a kind of cypress from the gum of which incense is made.

Phalūt (L).—The denuded peak, so called because the top of the mountain is bare of forest. The name has been corrupted by the Bhotiās to Phalilung, meaning the shield of the winds and apparently referring to the fact that the mountain helps to shield Lower Sikkim from the south-west monsoon. The mountain is accordingly called Phalalum in the Statistical Account of Bengal.

Phubsering (T).—The name is said to be that of a Bhotiā *sardār*, who first opened out the tea-garden now known by this designation. Properly it is Phurpusring, *i.e.*, Sring who was born on a Thursday (*phurpu*), it being a common practice to name Tibetan children after the day on which they are born.

Pobong (L).—The place of the *Po* bamboo.

Pulbazar (N).—The bridge market, the name of a village at the bridge over the Little Rangīt and of other places similarly situated.

Rammān (L).—A corruption of *Ra*, a surging and tumultuous advance, and *Mong*, a demon, the name of the lake Mong-do, beyond Phalūt, from which this river issues.

Rangarun (L).—The turning of the great river. For an explanation of this name, *see* the article above on Rangarun.

Rangfo (L).—The muddy brown extended water. A rivulet rising in the reddish laterite soil of the low outer hills, which joins the Tīsta near Sivok.

Rangīt (N).—The Nepāli corruption of Rangnyit, the river being known by the Bhotiās as the Rangnyit chhu, and by the Lepchās as the Rangnyit ung. The latter is the original name and means the two extended waters, *i. e.*, the Great and Little Rangīt, which are the principal rivers of Sikkim proper, as the Tīsta, which is larger, rises beyond Sikkim.

Rangjo (L).—The precipitous river, the name of a tributary of the Tīsta.

Rangli (L).—The Lepchā's house.

Rangliot (L).—The place of the receding waters. *See* the article on Rangarun.

Rangnu (L).—The straight-going river.

Rangpo (L).—The shifting or wandering river.

Rangtong (L).—The southern river.

Rayeng (L).—The spread-out or shallow river, a name given to a shallow and broadish stream near Sivok.

Rhenok (T).—The black hill. The soil of the hill is a very black humus. The name may however be a corruption of a word meaning the hill whose outline is like that of a nose.

Rikhyisum or **Risum**(B).—The three hills, so called because three ridges meet at this site.

Rillī (L).—The swirling river.

Rishi (T).—The hill top (*ri-tse*).

Rishihāt (N).—The place of the saint.

Rishi-lā (T).—The pass of the slipping mountain (*ri-shik*) so called because the pass has the appearance of a land slip.

Sabargūm (L).—The cliff of the musk-deer, a name given to the peak between Sandakphu and Phalūt because of the animals which used to haunt it.

Sandakphu (T).—The height of the poison-plant, *i.e.*, the aconite, which grows thickly for some distance below the peak known by this name.

Senchal (L).—The damp, misty hill, a name given to the ridge near Darjeeling, which overlooks the plains, receives the full force of the monsoon, and is cloud-capped for the greater part of the year.

Sepoydhura (N).—The sepoys' lines, the lines of the old Pioneer sepoys now disbanded.

Silīgurī.—The stony site. A Koch name due to the fact that the bed of the Mahānadī by which Silīgurī is situated is a mass of broken stone.

Simāna-bastī (N).—The boundary village, the name of a small village on the Nepāl frontier.

Singalīlā (L).—The mountain of the alder tree (*single-lā*).

Singtom (L).—An abbreviation of *shing*, a garden, and *tām*, a level spot.

Siniolchu (L).—The peak of the seven devils, Siniol being the name of a family of seven devils who fought there in bye-gone ages. It may be suggested, however, that the name is a corruption of *Sin* (mist), *sonon* (snow) and *chu* (mountain), *i.e.*, the mountain of mist and snow.

Sitong (L).—Tiger hill. A hill near Kurseong overlooking the Tarai, which was formerly frequented by tigers.

Sivok (L).—More properly Suvok, from *su*, a breeze or rush of cool air, and *vok*, concentrated. The mouth of the gorge whence the Tīsta debouches on the plains, along which a strong breeze blows.

Sonāda (L).—The bear's lair. Bears formerly frequented the neighbourhood and are still found there.

Sukiāpokhri (N).—The dry pool.

Suknā (N).—The dry site. Suknā is in the Tarai, but on a plateau at the base of a spur, where the water level is comparatively low.

Takvār (L).—A corruption of *tak*, a hook-thread, and *vor*, a fish-hook, a name suggested by the curve of the land.

Tanglu (L).—The hill of fir trees.

Tendong (L).—The uplifted horn. For an explanation of the meaning of this name, *see* the article on Darjeeling.

Tindhāria (N).—The three ridges.

Tirihāna (N).—A form of Tarai, *i.e.*, a swamp or marshy tract.

Tīsta (N).—An abbreviation of Trisrota, *i.e.*, the three currents. Until 1787 A. D. when it suddenly forsook its old bed and opened for itself a new channel, the Tīsta, on emerging from the hills, divided into three portions—the Atrai, Purnabhadra and Karatoyā, which each followed independent courses to the Ganges and Brahmaputra. The Bhotiā name for this river is Tsang-chhu or the pure water, while the Lepchās call it the Rangnyung or the great straight-going water, in allusion to the fact that it continues in a straight unaltered course in spite of receiving a great accession to its waters from the Great Rangit joining it at right angles. It has also been suggested, however, that the word is derived from *di* or *ti*, the Bodo word for water.

Tung (N).—The place of the *tun* tree (*Cedrela toona*).

INDEX.

A.

B.

E.

F.

N.

O.

P.

S.

T.

U

V.

W.

Y.

Z.